Monographs in

Number Theory

Volume 2

Topics in

Number Theory

Monographs in

Number Theory

Volume 2

Topics in
Number Theory

Minking Eie
National Chung Cheng University, Taiwan

World Scientific

NEW JERSEY · LONDON · SINGAPORE · BEIJING · SHANGHAI · HONG KONG · TAIPEI · CHENNAI

Published by

World Scientific Publishing Co. Pte. Ltd.
5 Toh Tuck Link, Singapore 596224
USA office: 27 Warren Street, Suite 401-402, Hackensack, NJ 07601
UK office: 57 Shelton Street, Covent Garden, London WC2H 9HE

British Library Cataloguing-in-Publication Data
A catalogue record for this book is available from the British Library.

TOPICS IN NUMBER THEORY
Monographs in Number Theory — Vol. 2
Copyright © 2009 by World Scientific Publishing Co. Pte. Ltd.

ISBN-13 978-981-283-518-5
ISBN-10 981-283-518-0

Printed in Singapore.

Preface

Since I left the Institute of Mathematics at Academia Sinica in 1991, I have conducted a graduate course on number theory every year at the National Chung Cheng University. The title of the course is usually "Special Topics in Number Theory" or "Theory of Modular Forms of Several Variables". Unfortunately, there are no textbooks available for such a course and I had to write my own lecture notes, lifting materials from books by T. Shimura [48], E. Freitag [31], W. L. Baily, Jr. [6] and T. Miyake [42]. To enrich the contents of the lecture notes, I have included some of the results from my own research work.

In such a course as mentioned above, we usually begin with the theory of modular forms of one variable, followed by the theory of zeta functions, new approaches to Bernoulli identities and Kummer Congruences, the theory of modular forms of several variables, the theory of Jacobi forms over Cayley numbers and finally the theory of Jacobi forms of several variables over Cayley numbers. The theory of Jacobi forms is based on my research papers from 1991 to 2000. Students are encouraged to conduct research related to the materials mentioned above. Eventually, a total of 6 Ph.D. students and more than 30 M.Sc. students graduated under this programme.

The theory of modular forms of several variables [31] was initiated by C. L. Siegel around 1945 in an attempt to generalize the theory of modular forms of one variable. The theory was employed by M. Eichler and D. Zagier [15] in 1985 to give a definition of Jacobi forms. Indeed the coefficients of the Fourier-Jacobi expansions of modular forms of rank two are the main sources of Jacobi forms. It is still a difficult problem to construct Jacobi forms besides the Eisenstein series. Even the Eisenstein series of Jacobi forms of several variables are difficult to construct. All we can say is that many results concerning modular forms of several variables remain to be studied in depth.

Georg-August-Universität Göttingen is the center of development of modular forms with masters like Siegel and U. Christian. I spent a year at Georg-August-Universität Göttingen as a Sonderforschungsbereich (special research fellow) in 1985, three years after my graduation from the University of Chicago in 1982.

It is almost 60 years since the birth of modular forms of several variables in Germany. It is an appropriate time to have a textbook on the theory in English rather than in German. This is my main motivation for writing this book.

Minking Eie
April 19, 2008

Glossary

Contents

Part I

Theory of Modular Forms
of One Variable

Chapter 1

Group Actions of the Modular Group

Theory of modular forms begins with the group action of the special linear group $SL_2(\mathbb{R})$ on the upper half-plane $\mathcal{H}$. For a nonnegative integer k, a holomorphic function f defined on the upper half-plane is a *modular form of weight k* if

$$f\left(\frac{az+b}{cz+d}\right) = (cz+d)^k f(z)$$

for all $\left[\begin{smallmatrix} a & b \\ c & d \end{smallmatrix}\right]$ in the modular group $SL_2(\mathbb{Z})$ and $f(z)$ has a Fourier expanion

$$f(z) = \sum_{n=0}^{\infty} a_n e^{2\pi i n z}.$$

The vector space of modular forms has finite dimension and its dimension can be computed by Selberg trace formula.

1.1. The Upper Half-Plane

Let $\mathcal{H}$ be the upper half-plane of the complex plane defined by

$$\mathcal{H} = \{z \in \mathbb{C} \mid \operatorname{Im} z > 0\}$$

3

and $SL_2(\mathbb{R})$ be the special linear group over $\mathbb{R}$ given by

$$SL_2(\mathbb{R}) = \left\{ \begin{bmatrix} a & b \\ c & d \end{bmatrix} \,\middle|\, a, b, c, d \in \mathbb{R}, ad - bc = 1 \right\}.$$

The special linear group $SL_2(\mathbb{R})$ acts on $\mathscr{H}$ by the action

$$g = \begin{bmatrix} a & b \\ c & d \end{bmatrix} : z \mapsto g(z) = \frac{az + b}{cz + d}.$$

Indeed, one has

$$\operatorname{Im} g(z) = \frac{1}{2i} \left(\frac{az + b}{cz + d} - \frac{a\bar{z} + b}{c\bar{z} + d} \right) = \frac{1}{2i} \frac{(ad - bc)(z - \bar{z})}{|cz + d|^2} = \frac{\operatorname{Im} z}{|cz + d|^2}.$$

Here we have the following facts about the action.

Fact 1. The action is transitive in the sense that for z_1, $z_2 \in \mathscr{H}$, there exists $g \in SL_2(\mathbb{R})$ such that $g(z_1) = z_2$. It suffices to consider the special case when $z_1 = i$ and $z_2 = x + iy$. If we let

$$M = \begin{bmatrix} 1 & x \\ 0 & 1 \end{bmatrix} \begin{bmatrix} \sqrt{y} & 0 \\ 0 & 1/\sqrt{y} \end{bmatrix},$$

then $M \in SL_2(\mathbb{R})$ and $M(i) = x + iy$.

Fact 2. When a group G acts on a set S, the isotropy subgroup of G at a particular point x in S is defined as

$$\{ g \in G \mid g(x) = x \}.$$

The isotropy subgroup of $G = SL_2(\mathbb{R})$ at $z = i$ is given by

$$K = \left\{ \begin{bmatrix} \cos\theta & \sin\theta \\ -\sin\theta & \cos\theta \end{bmatrix} \,\middle|\, \theta \in \mathbb{R} \right\} \cong S^1 = \left\{ e^{i\theta} \mid \theta \in \mathbb{R} \right\}.$$

It follows from the fact that

$$\frac{ai + b}{ci + d} = i \iff a = d, \quad b = -c$$

and $ad - bc = 1$ implies $a^2 + b^2 = 1$. There is a one-to-one correspondence between the left cosets of K in G and points in $\mathscr{H}$ given by

$$gK \rightsquigarrow g(i).$$

So topologically, we have $G/K \cong \mathscr{H}$.

Fact 3. For each $g \in \mathrm{SL}_2(\mathbb{R})$, g has the *Iwasawa decomposition* [42]

$$g = \begin{bmatrix} 1 & x \\ 0 & 1 \end{bmatrix} \begin{bmatrix} \sqrt{y} & 0 \\ 0 & 1/\sqrt{y} \end{bmatrix} \begin{bmatrix} \cos\theta & \sin\theta \\ -\sin\theta & \cos\theta \end{bmatrix}.$$

Suppose that $g(i) = x + iy$ and the matrix M is given as in Fact 1. Then

$$g(i) = x + iy = M(i)$$

and hence $M^{-1}g \in K$. So there exists a $P \in K$ such that $g = MP$ and g has the decomposition as asserted.

Fact 4. For all $g \in G$, we have

$$\frac{dg(z)}{dz} = \frac{1}{(cz+d)^2}.$$

Set $j(g, z) = cz + d$ if $g = \begin{bmatrix} a & b \\ c & d \end{bmatrix} \in \mathrm{SL}_2(\mathbb{R})$. Then the chain rule on differentiation implies the following *cocycle condition*

$$j(g_1 g_2, z) = j(g_1, g_2(z)) j(g_2, z). \tag{1.1.1}$$

1.2. The Modular Group

Let $G = \mathrm{SL}_2(\mathbb{R})$ and $\overline{G} = \mathrm{SL}_2(\mathbb{R})/\{\pm E_2\}$ be the project special linear group, where E_2 is the 2×2 identity matrix $\begin{bmatrix} 1 & 0 \\ 0 & 1 \end{bmatrix}$. Then

$$\overline{G} \cong \mathrm{Aut}(\mathscr{H}),$$

the set of one-to-one and bi-holomorphic functions from $\mathscr{H}$ onto itself. Write Γ for $\mathrm{SL}_2(\mathbb{Z})$, which is $\mathrm{SL}_2(\mathbb{R}) \cap M_2(\mathbb{Z})$. Set

$$\overline{\Gamma} = \mathrm{SL}_2(\mathbb{Z})/\{\pm E_2\}$$

and

$$\Gamma_0 = \left\{ \pm \begin{bmatrix} 1 & n \\ 0 & 1 \end{bmatrix} \ \middle| \ n \in \mathbb{Z} \right\}.$$

The set $\overline{\Gamma}$ is called the *full modular group* and Γ_0 is called the *maximal parabolic subgroup of* Γ.

Proposition 1.2.1. *There is a one-to-one correspondence between the set of ordered pairs of integers (c, d) such that $\gcd(c, d) = 1$ and the set of right cosets of Γ_0 in Γ.*

Proof. Given a pair of integers (c, d) with $\gcd(c, d) = 1$, there exist two integers a and b such that $ad - bc = 1$, so that

$$M = \begin{bmatrix} a & b \\ c & d \end{bmatrix} \in \Gamma.$$

Also if

$$M' = \begin{bmatrix} * & * \\ c & d \end{bmatrix} \in \Gamma$$

has the same c, d block as M, then

$$M'M^{-1} = \begin{bmatrix} * & * \\ c & d \end{bmatrix} \begin{bmatrix} d & -b \\ -c & a \end{bmatrix} = \pm \begin{bmatrix} 1 & * \\ 0 & 1 \end{bmatrix} \in \Gamma_0.$$

It follows $M' \in \Gamma_0 M$ and hence $\Gamma_0 M' = \Gamma_0 M$. $\qquad\square$

A domain $\mathscr{F}$ on $\mathscr{H}$ is called a *fundamental domain* of $\mathscr{H}$ with respect to $\Gamma = \mathrm{SL}_2(\mathbb{Z})$ if it satisfies the following conditions:

(a) For all $z \in \mathscr{H}$, there exists $g \in \Gamma$ such that $g(z) \in \mathscr{F}$;

(b) If z_1 and z_2 are two distinct elements in $\mathscr{F}$, and $g \in \Gamma \setminus \{\pm E_2\}$ such that $g(z_1) = z_2$, then $z_1, z_2 \in \partial\mathscr{F}$, the boundary of $\mathscr{F}$.

Roughly speaking, a fundamental domain is a set of representatives for the orbit space $\Gamma \setminus \mathscr{H}$ with possible exceptions for its boundary points. Two points z_1 and z_2 lie in the same orbit if and only if $z_1 = g(z_2)$ for some $g \in \Gamma$.

Proposition 1.2.2. *A fundamental domain of $\mathscr{H}$ with respect to $\Gamma = \mathrm{SL}_2(\mathbb{Z})$ is given by*

$$\mathscr{F} = \left\{ z \in \mathscr{H} \;\middle|\; |z| \geq 1, -\frac{1}{2} \leq \mathrm{Re}\, z \leq \frac{1}{2} \right\}.$$

Here we describe the general procedure to obtain the fundamental domain.

Step I. Given $z \in \mathscr{H}$, the set

$$S = \{(c,d) \mid c,d \in \mathbb{Z}, \gcd(c,d) = 1, |cz + d| \leq 1\}$$

is finite. So we can pick up (c,d) in S so that $|cz + d|$ is minimal. Let

$$g = \begin{bmatrix} a & b \\ c & d \end{bmatrix} \in \Gamma \quad \text{and} \quad z_1 = g(z).$$

Then

$$\operatorname{Im} z_1 = \frac{\operatorname{Im} z}{|cz + d|^2}$$

is maximal in the set

$$\{g(z) \mid g \in \Gamma\}.$$

By applying a translation $T_b \colon z \mapsto z + b$ with a suitable $b \in \mathbb{Z}$, we may assume that $z_2 = T_b(z_1)$ satisfies the condition

$$-\frac{1}{2} \leq \operatorname{Re} z_2 \leq \frac{1}{2}.$$

We claim that $z_2 \in \mathscr{F}$. Suppose to the contrary that $|z_2| < 1$ and $-z_2^{-1} \in \mathscr{H}$. This implies that

$$\operatorname{Im}(-z_2^{-1}) = \frac{\operatorname{Im} z_2}{|z_2|^2} > \operatorname{Im} z_2.$$

This contradicts the fact that $\operatorname{Im} z_2$ is maximal in the set $\{g(z) \mid g \in \Gamma\}$.

Step II. Let $z \in \mathscr{F}$ and $g = \begin{bmatrix} a & b \\ c & d \end{bmatrix} \in \Gamma$ such that $g(z) \in \mathscr{F}$. In particular, we have

$$|cz + d| = 1 \quad \text{or} \quad (cx + d)^2 + c^2 y^2 = 1,$$

where $z = x + iy$. As $y \geq \sqrt{3}/2$ in the fundamental domain, we have $c = 0$, 1 or -1.

(A) If $c = 0$, then $d = \pm 1$, g is a translation and $g(z) = z \pm 1$. So it forces that $\operatorname{Re} z = \pm 1/2$ and z is in the boundary of $\mathscr{F}$.

(B) If $c = 1$, then $|z + d| = 1$. This means that the distance between z and points with integral coordinates in the real axis is one. Therefore, we have

(1) $d = 0$ except $z = e^{2\pi i/3}$ ($d = 0$ or 1) or $z = e^{\pi i/3}$ ($d = 0$ or -1).

(2) $d = 0$, then $|z| = 1$, $g(z) = -1/z + a$, where $a = 0$ except $z = e^{2\pi i/3}$ or $z = e^{\pi i/3}$ and $a = 0$, $z = i$.

Remark 1.2.3. The isotropy subgroup of Γ at $z \in \mathscr{F}$ is the trivial subgroup $\{\pm E_2\}$ except $z = i$ or $z = e^{2\pi i/3}$ or $z = e^{\pi i/3}$.

The isotropy subgroup of Γ at $z = i$ is

$$\Gamma \cap K = \langle S \rangle, \quad S = \begin{bmatrix} 0 & -1 \\ 1 & 0 \end{bmatrix}$$

which is a subgroup of Γ of order 4.

The isotropy subgroup of Γ at $z = e^{2\pi i/3} = \omega$ can be obtained by a direct calculation as follows:

$$\frac{a\omega + b}{c\omega + d} = \omega, \quad a\omega + b = c\omega^2 + d\omega = -c(1 + \omega) + d\omega.$$

It follows $a = d - c$, $b = -c$ and

$$\begin{bmatrix} d - c & -c \\ c & d \end{bmatrix} \in \Gamma \quad \text{with} \quad c^2 - cd + d^2 = 1.$$

The solutions to the equation are

$$(c, d) = (1, 0), (-1, 0), (1, 1), (-1, -1), (0, 1), (0, -1).$$

The isotropy subgroup is a cyclic group of order 6 generated by

$$ST = \begin{bmatrix} 0 & -1 \\ 1 & 0 \end{bmatrix} \begin{bmatrix} 1 & 1 \\ 0 & 1 \end{bmatrix} = \begin{bmatrix} 0 & -1 \\ 1 & 1 \end{bmatrix}.$$

Proposition 1.2.4. *The group* $\Gamma = \mathrm{SL}_2(\mathbb{Z})$ *is generated by*

$$S = \begin{bmatrix} 0 & -1 \\ 1 & 0 \end{bmatrix} \quad \text{and} \quad T = \begin{bmatrix} 1 & 1 \\ 0 & 1 \end{bmatrix}.$$

Proof. Let G be the subgroup of Γ generated by S and T. Suppose $G \neq \Gamma$. Let

$$c_0 = \min \left\{ |c| \, \middle| \, \begin{bmatrix} a & b \\ c & d \end{bmatrix} \in \Gamma \setminus G \right\}.$$

Note that $c_0 > 0$ since G contains Γ_0. Let

$$M = \begin{bmatrix} a_0 & b_0 \\ c_0 & d_0 \end{bmatrix} \in \Gamma \setminus G.$$

Then there exist integers q and r such that

$$a_0 = c_0 q + r, \quad 0 \leq r < c_0.$$

Since

$$T^{-q} M = \begin{bmatrix} a_0 - c_0 q & b_0 - d_0 q \\ c_0 & d_0 \end{bmatrix} = \begin{bmatrix} r & b_0 - d_0 q \\ c_0 & d_0 \end{bmatrix},$$

by minimality of c_0, we conclude that

$$ST^{-q} M = \begin{bmatrix} -c_0 & -d_0 \\ r & b_0 - d_0 q \end{bmatrix} \in G.$$

It follows $M \in G$. This contradicts the fact that $M \in \Gamma \setminus G$ and we conclude that $\Gamma = G$. $\qquad \square$

1.3. Modular Forms

Let k be an integer. A holomorphic function $f \colon \mathscr{H} \to \mathbb{C}$ is called a *modular form of weight k with respect to the full modular group* $\Gamma = \mathrm{SL}_2(\mathbb{Z})$ if it satisfies the following conditions:

(M-1) $f\left(\dfrac{az+b}{cz+d}\right) = (cz+d)^k f(z)$ for all $\begin{bmatrix} a & b \\ c & d \end{bmatrix} \in \Gamma$;

(M-2) f has a Fourier expansion of the form

$$f(z) = \sum_{n=0}^{\infty} a_n e^{2\pi i n z}.$$

A modular form f is a *cusp form* if it satisfies the further condition that $a_0 = 0$.

Remark 1.3.1. When $\begin{bmatrix} a & b \\ c & d \end{bmatrix} = \begin{bmatrix} -1 & 0 \\ 0 & -1 \end{bmatrix}$, we have

$$f(z) = (-1)^k f(z).$$

So k must be even, otherwise $f(z) = 0$. In other words, there is no nonzero modular forms of odd weights.

Remark 1.3.2. Since $\Gamma = \mathrm{SL}_2(\mathbb{Z})$ is generated by

$$S = \begin{bmatrix} 0 & -1 \\ 1 & 0 \end{bmatrix} \quad \text{and} \quad T = \begin{bmatrix} 1 & 1 \\ 0 & 1 \end{bmatrix}.$$

The first condition can be replaced by

$$f(z+1) = f(z) \quad \text{and} \quad f\left(-\frac{1}{z}\right) = z^k f(z).$$

A typical example of modular forms is given by the *Eisenstein series* defined by

$$G_k(z) = \sum_{(m,n)\neq(0,0)} \frac{1}{(mz+n)^k}, \quad k \text{ even}, \quad k \geq 4.$$

First we test the convergence of the infinite series used to define the functions. Set

$$d = \min\{1, |z|\}.$$

It is the minimal distance from the origin to the parallelogram with vertices $z+1$, $z-1$, $-z+1$ and $-z-1$. Then

$$\sum_{(m,n)\neq(0,0)} \frac{1}{|mz+n|^k} \leq \sum_{n=1}^{\infty} \frac{8n}{(nd)^k} = \frac{8}{d^k} \sum_{n=1}^{\infty} \frac{1}{n^{k-1}}.$$

Thus the series is absolutely convergent on any compact subset of $\mathcal{H}$ when $k > 2$.

Next, we prove that $G_k(z)$ is indeed a modular form of weight k. Let

$$\mathscr{L} = \{mz + n \mid m, n \in \mathbb{Z}\}.$$

Then $\mathscr{L}$ is a lattice of $\mathbb{C}$ and $G_k(z)$ can be rewritten as

$$G_k(z) = \sum_{\substack{\lambda \in \mathscr{L} \\ \lambda \neq 0}} \lambda^{-k}.$$

For each $\begin{bmatrix} a & b \\ c & d \end{bmatrix} \in \Gamma$, the mapping

$$z \mapsto az + b, \quad 1 \mapsto cz + d$$

extends to a one-to-one linear mapping from $\mathscr{L}$ onto $\mathscr{L}$ and hence

$$
\begin{aligned}
G_k\left(\frac{az+b}{cz+d}\right) &= (cz+d)^k \sum_{(m,n)\neq(0,0)} [m(az+b)+n(cz+d)]^{-k} \\
&= (cz+d)^k \sum_{\substack{\lambda\in\mathscr{L} \\ \lambda\neq 0}} \lambda^{-k} \\
&= (cz+d)^k G_k(z).
\end{aligned}
$$

It remains to find the Fourier expansion of $G_k(z)$. Rewrite $G_k(z)$ as

$$
G_k(z) = 2\zeta(k) + 2\sum_{m=1}^{\infty}\sum_{n\in\mathbb{Z}}(mz+n)^{-k},
$$

where $\zeta(s)$ is the Riemann zeta function defined to be

$$
\zeta(s) = \sum_{n=1}^{\infty}\frac{1}{n^s}, \quad \operatorname{Re} s > 1.
$$

Proposition 1.3.3. *For any even integer $k \geq 2$, one has*

$$
\sum_{n\in\mathbb{Z}}\frac{1}{(z+n)^k} = \frac{(-2\pi i)^k}{(k-1)!}\sum_{n=1}^{\infty}n^{k-1}e^{2\pi i n z}.
$$

Proof. Let

$$
f(x) = \sum_{n\in\mathbb{Z}}\frac{1}{(x+iy+n)^k}.
$$

Then $f(x+1) = f(x)$ and $f(x)$ is a periodic function, so it has a Fourier expansion

$$
f(x) = \sum_{n=-\infty}^{\infty} a_n(y)e^{2\pi i n x}
$$

with

$$
\begin{aligned}
a_n(y) &= \int_0^1 f(x)e^{-2\pi i n x}\,dx \\
&= \int_0^1 \sum_{m\in\mathbb{Z}}(x+iy+m)^{-k}e^{-2\pi i n x}\,dx \\
&= \int_{-\infty}^{\infty}\frac{e^{-2\pi i n x}}{(x+iy)^k}\,dx \\
&= \begin{cases} 0, & \text{if } n \leq 0; \\ \dfrac{(-2\pi i)^k}{(k-1)!}n^{k-1}e^{-2\pi n y}, & \text{if } n > 0. \end{cases}
\end{aligned}
$$

See Ahlfors [1] for the details of the above evaluation through contour integrals.   □

Proposition 1.3.4. *For even integer $k \geq 4$, $G_k(z)$ is a modular form of weight k with a Fourier expansion of the form*

$$G_k(z) = 2\zeta(k) \left\{ 1 - \frac{2k}{B_k} \sum_{n=1}^{\infty} \sigma_{k-1}(n) e^{2\pi i n z} \right\}.$$

Proof. Applying the preceding proposition, we have

$$G_k(z) = 2\zeta(k) + \frac{2(-2\pi i)^k}{(k-1)!} \sum_{n=1}^{\infty} \sum_{m=1}^{\infty} n^{k-1} e^{2\pi i m n z}.$$

Note that (see, for example, Apostol [3, p. 266])

$$\zeta(k) = \frac{(-1)^{k/2-1}(2\pi)^k B_k}{2(k!)},$$

where B_k is the kth Bernoulli number defined by

$$\frac{t}{e^t - 1} = \sum_{k=0}^{\infty} \frac{B_k t^k}{k!}, \quad |t| < 2\pi.$$

It follows that

$$G_k(z) = 2\zeta(k) \left\{ 1 - \frac{2k}{B_k} \sum_{n=1}^{\infty} \sigma_{k-1}(n) e^{2\pi i n z} \right\},$$

where $\sigma_k(n)$ is the divisor function given by

$$\sigma_k(n) = \sum_{d|n} d^k.$$ □

Here we provide the second proof for the transformation formula of $G_k(z)$ under the modular group. We have

$$G_k(z) = \sum_{(m,n) \neq (0,0)} (mz+n)^{-k} = \zeta(k) \sum_{(c,d)=1} (cz+d)^{-k}$$

$$= 2\zeta(k) \sum_{\gamma \in \Gamma/\Gamma_0} j(\gamma, z)^{-k}.$$

Here we use the elementary fact that every pair of integers m and n with $(m, n) \neq (0, 0)$ can be written uniquely as $m = kc$, $n = kd$ with k a positive integer and c, d being relatively prime integers. It follows for $M = \begin{bmatrix} a & b \\ c & d \end{bmatrix} \in \Gamma$ that

$$\begin{aligned} G_k(M(z)) &= 2\zeta(k) \sum_{\gamma \in \Gamma/\Gamma_0} j(\gamma, M(z))^{-k} \\ &= j(M, z)^k 2\zeta(k) \sum_{\gamma \in \Gamma/\Gamma_0} j(\gamma M, z)^{-k} \\ &= j(M, z)^k G_k(z) \end{aligned}$$

if we employ the cocycle condition (1.1.1)

$$j(\gamma M, z) = j(\gamma, M(z)) j(M, z).$$

The *normalized Eisenstein series* $E_k(z)$ is the quotient of $G_k(z)$ by $2\zeta(k)$. Thus

$$E_k(z) = \sum_{\gamma \in \Gamma/\Gamma_0} j(\gamma, z)^{-k} = 1 - \frac{2k}{B_k} \sum_{n=1}^{\infty} \sigma_{k-1}(n) e^{2\pi i n z}.$$

In particular, we have

$$E_4(z) = 1 + 240 \sum_{n=1}^{\infty} \sigma_3(n) e^{2\pi i n z} \quad \text{and} \quad E_6(z) = 1 - 504 \sum_{n=1}^{\infty} \sigma_5(n) e^{2\pi i n z}.$$

Remark 1.3.5. It is well-known that

$$\sin z = z \prod_{n=1}^{\infty} \left(1 - \frac{z^2}{n^2 \pi^2} \right).$$

Taking the logarithmic derivative, we get

$$z \cot z = 1 + 2 \sum_{n=1}^{\infty} \frac{z^2}{z^2 - n^2 \pi^2} = 1 - 2 \sum_{n=1}^{\infty} \sum_{k=1}^{\infty} \frac{z^{2k}}{n^{2k} \pi^{2k}} = 1 - 2 \sum_{k=1}^{\infty} \zeta(2k) \frac{z^{2k}}{\pi^{2k}}.$$

On the other hand, we also have

$$z \cot z = z \frac{\cos z}{\sin z} = iz \frac{e^{2iz} + 1}{e^{2iz} - 1} = iz + \frac{2iz}{e^{2iz} - 1} = 1 + \sum_{k=1}^{\infty} \frac{B_{2k}(2iz)^{2k}}{(2k)!}.$$

Comparing the coefficients of z^{2k} on two power series expansions of $z \cot z$, we get

$$\zeta(2k) = \frac{(-1)^{k-1} B_{2k}(2\pi)^{2k}}{2(2k)!}.$$

1.4. Exercises

1. Let B_n, $n = 0, 1, 2, \ldots$, be Bernoulli numbers defined by the generating function

$$\frac{t}{e^t - 1} = \sum_{k=0}^{\infty} \frac{B_k t^k}{k!}, \quad |t| < 2\pi.$$

Prove that $B_0 = 1$ and for $n \geq 2$,

$$\binom{n}{n-1} B_{n-1} + \binom{n}{n-2} B_{n-2} + \cdots + \binom{n}{0} B_0 = 0.$$

Determine the values of $B_1, B_2, B_3, B_4, B_5, B_6$ by the above recursive formula.

2. For positive integers m and N, let

$$S_m(N) = \sum_{k=1}^{N-1} k^m.$$

Prove that

$$S_m(N) = \frac{1}{m+1} \sum_{k=0}^{m} \binom{m+1}{k} B_k N^{m+1-k}.$$

3. Prove Wallis's product formula

$$\frac{\pi}{2} = \frac{2 \cdot 2}{1 \cdot 3} \cdot \frac{4 \cdot 4}{3 \cdot 5} \cdots \frac{2m \cdot 2m}{(2m-1)(2m+1)} \cdots.$$

[*Hint.* Use the product formula for $\sin z$ at $z = \pi/2$.]

4. Prove the product formula for $\sin z$,

$$\sin z = z \prod_{n=1}^{\infty} \left(1 - \frac{z^2}{n^2 \pi^2}\right).$$

5. Give the details of the proof for $y > 0$ that

$$\int_{-\infty}^{\infty} \frac{e^{-2\pi i n x}}{(x + iy)^k} \, dx = \begin{cases} 0, & \text{if } n \leq 0; \\ \dfrac{(-2\pi i)^k}{(k-1)!} n^{k-1} e^{-2\pi n y}, & \text{if } n > 0. \end{cases}$$

Chapter 2

The Gamma and Zeta Functions

2.1. A First Look at the Riemann Zeta Function and the Gamma Function

The well-known *Riemann zeta function* $\zeta(s)$ is defined by

$$\zeta(s) = \sum_{n=1}^{\infty} n^{-s}, \quad \operatorname{Re} s > 1.$$

Let $s = \sigma + it$ with $\sigma, t \in \mathbb{R}$. Then

$$|n^{-s}| = |\exp(-(\sigma + it)) \log n| = \exp(-\sigma \log n) = n^{-\sigma}.$$

Thus the infinite series is dominated term by term by the series

$$\sum_{n=1}^{\infty} n^{-\sigma}$$

which is convergent for $\sigma > 1$. Hence $\zeta(s)$ is analytic as a function in s for $\operatorname{Re} s = \sigma > 1$. Furthermore, by Fundamental Theorem of Arithmetic, we have the infinite product formula (Euler product)

$$\zeta(s) = \prod_p \left(1 + p^{-s} + p^{-2s} + \cdots + p^{-ks} + \cdots\right)$$

$$= \prod_p (1 - p^{-s})^{-1}.$$

Thus $\zeta(s)$ has no zero for $\operatorname{Re} s > 1$ since its factor $(1 - p^{-s})^{-1}$ is never zero. The *gamma function* $\Gamma(s)$ (or the *factorial function*) is defined as

$$\Gamma(s) = \int_0^\infty t^{s-1} e^{-t}\, dt, \quad \operatorname{Re} s > 0.$$

In light of its *functional equation* (see, for example, Apostol [3, p. 260])

$$\pi^{-s/2}\Gamma\left(\frac{s}{2}\right)\zeta(s) = \pi^{-(1-s)/2}\Gamma\left(\frac{1-s}{2}\right)\zeta(1-s) \qquad (2.1.1)$$

which we would discuss later, so $\zeta(s)$ also has no zero for $\operatorname{Re} s < 0$.

Here $\zeta(-2m)$, $m = 1, 2, 3, \ldots$, must be zero since the gamma function $\Gamma(s/2)$ has a simple pole at $s = -2m$ as we shall see. The well-known Riemann Hypothesis then asserted that all the nontrivial zeros lie on the critical line $\operatorname{Re} s = 1/2$. Until now, this conjecture is still not resolved. Here we give the conjecture.

Proposition 2.1.1 (Riemann Hypothesis). *All the nontrivial zeros of* $\zeta(s)$ *lie on the critical line* $\operatorname{Re} s = 1/2$.

Remark 2.1.2. Let $\pi(x)$ be the number of prime numbers not exceeding x. The prime number theorem states that

$$\pi(x) \sim \frac{x}{\log x} \quad \text{as } x \to \infty,$$

or equivalently

$$\lim_{x \to \infty} \frac{\pi(x) \log x}{x} = 1.$$

The prime number theorem was proved in 1896 by J. Hadamard [33] and C. J. de la Vallée-Poussin [50] independently. If $N(T)$ denotes the number of zeros of $\zeta(s)$ inside the rectangle

$$0 < \operatorname{Re} s < 1, \quad 0 < \operatorname{Im} s < T,$$

then Littlewood [40] in 1914 gave

$$N(T) = \frac{1}{2\pi}T \log T - \frac{1 + \log 2\pi}{2\pi}T + o(\log T)$$

under the assumption of Riemann Hypothesis.

D. Hilbert remarked in his lecture at the Paris Congress that Riemann Hypothesis is equivalent to

$$\pi(x) = \int_2^x \frac{dx}{\log x} + O(\sqrt{x}\log x) \quad \text{as } x \to \infty.$$

It is also equivalent to

$$\sum_{n=1}^N \mu(n) = O(N^{1/2+\epsilon}) \quad \text{as } N \to \infty$$

for any $\epsilon > 0$, where $\mu(n)$ is the Möbius function defined by

$$\mu(n) = \begin{cases} 1, & \text{if } n = 1; \\ (-1)^k, & \text{if } n = p_1 p_2 \cdots p_k \text{ and } p_i \neq p_j \text{ for } i \neq j; \\ 0, & \text{otherwise.} \end{cases}$$

Recall the gamma function $\Gamma(s)$ is defined as

$$\Gamma(s) = \int_0^\infty t^{s-1} e^{-t}\, dt, \quad \text{Re}\, s > 0.$$

Right from the above definition, we have the following functional equation.

Proposition 2.1.3. *For* $\text{Re}\, s > 0$, *one has*

$$\Gamma(s+1) = s\Gamma(s).$$

Proof. This follows immediately from the following computation:

$$\Gamma(s+1) = \int_0^\infty t^s e^{-t}\, dt = -t^s e^{-t}\Big|_0^\infty + \int_0^\infty e^{-t}\, dt^s$$

$$= 0 + s\int_0^\infty t^{s-1} e^{-t}\, dt = s\Gamma(s). \qquad \square$$

With the above functional equation, we are able to extend the domain of definition of $\Gamma(s)$ to the whole complex plane. For $s \in \mathbb{C}$ with $\operatorname{Re} s > -m$, we define

$$\Gamma(s) = \frac{\Gamma(s+m)}{s(s+1)\cdots(s+m-1)}.$$

So we have the following properties for the gamma function $\Gamma(s)$.

1. The function $\Gamma(s)$ has an analytic continuation in the whole complex plane except for simple poles at $s = -m$, $m = 0, 1, 2, \ldots$, with the residue

$$\frac{(-1)^m}{m!}.$$

2. The function $\Gamma(s)$ satisfies the functional equation

$$\Gamma(s+1) = s\Gamma(s).$$

In particular, for any nonnegative integer n, we have $\Gamma(n+1) = n!$.

3. Weierstrass defined the gamma function as an infinite product (see, for example, Rademacher [45, p. 32])

$$\frac{1}{\Gamma(s)} = se^{\gamma s} \prod_{n=1}^{\infty} \left(1 + \frac{s}{n}\right) e^{-s/n},$$

where γ is the Euler constant given by

$$\gamma = \lim_{n \to \infty} \left(\sum_{k=1}^{n} \frac{1}{k} - \log n\right).$$

4. (Reflection formula, see, for example, Rademacher [45, p. 33]). We have $\Gamma(s)\Gamma(1-s) = \pi/\sin(\pi s)$. In particular, $\Gamma(1/2) = \sqrt{\pi}$. The above formula follows from the infinite product expression for $1/\Gamma(s)$ and

$$\sin(\pi s) = \pi s \prod_{n=1}^{\infty} \left(1 - \frac{s^2}{n^2}\right).$$

5. (Legendre's duplication formula, see, for example, Ahlfors [1, p. 200]). We have $\Gamma(s)\Gamma(s+1/2) = 2^{1-2s}\Gamma(1/2)\Gamma(2s)$.

2.2. Analytic Continuation of Riemann Zeta Function

A first step to give the analytic continuation of $\zeta(s)$ is to express $\zeta(s)\Gamma(s)$ as an improper integral when $\operatorname{Re} s > 1$. It is performed in the following proposition.

Proposition 2.2.1. *For* $\operatorname{Re} s > 1$, *one has*

$$\zeta(s)\Gamma(s) = \int_0^\infty \frac{t^{s-1}\, dt}{e^t - 1}.$$

Proof. We begin with the integral expression of the gamma function

$$\Gamma(s) = \int_0^\infty t^{s-1} e^{-t}\, dt, \quad \operatorname{Re} s > 0.$$

Setting $t = nt'$ and still write t in place of t', we get

$$n^{-s}\Gamma(s) = \int_0^\infty t^{s-1} e^{-nt}\, dt.$$

Summing together for all positive integers $n = 1, 2, 3, \ldots$, we get for $\operatorname{Re} s > 1$ that

$$\zeta(s)\Gamma(s) = \sum_{n=1}^\infty \int_0^\infty t^{s-1} e^{-nt}\, dt$$

$$= \int_0^\infty t^{s-1} \sum_{n=1}^\infty e^{-nt}\, dt$$

$$= \int_0^\infty \frac{t^{s-1}}{e^t - 1}\, dt.$$

In the above, we exchange the order of summation and integration which is justified by the well-known Lebesgue's dominated convergence theorem (LDCT) which appears in general textbooks of real analysis. $\qquad\square$

Let $L(\epsilon)$ be the "key" contour in the complex plane, consisting of the interval $[\epsilon, +\infty)$ twice in opposite direction and the circle $|z| = \epsilon$ in counterclockwise direction.

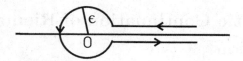

Figure 2.1: Key contour in the complex plane.

Proposition 2.2.2. *For* $\operatorname{Re} s > 1$ *and* $s \notin \mathbb{Z}$, *one has*

$$\zeta(s)\Gamma(s) = (e^{2\pi i s} - 1)^{-1} \int_{L(\epsilon)} \frac{z^{s-1} \, dz}{e^z - 1}.$$

Proof. For $\operatorname{Re} s > 1$, one has

$$\int_{L(\epsilon)} \frac{z^{s-1} \, dz}{e^z - 1} = -\int_{\epsilon}^{\infty} \frac{t^{s-1} \, dt}{e^t - 1} + \int_{|z|=\epsilon} \frac{z^{s-1} \, dz}{e^z - 1} + e^{2\pi i s} \int_{\epsilon}^{\infty} \frac{t^{s-1} \, dt}{e^t - 1}$$

$$= (e^{2\pi i s} - 1) \int_{\epsilon}^{\infty} \frac{t^{s-1} \, dt}{e^t - 1} + \int_{|z|=\epsilon} \frac{z^{s-1} \, dz}{e^z - 1}.$$

For $\operatorname{Re} s = \sigma > 1$, we have

$$\left| \int_{|z|=\epsilon} \frac{z^{s-1} \, dz}{e^z - 1} \right| \leq \int_{|z|=\epsilon} \left| \frac{z}{e^z - 1} \right| |z^{s-1}| \frac{|dz|}{|z|}$$

$$\leq M\epsilon^{\sigma - 1}$$

which approaches 0 as $\epsilon \to 0$. It follows that for $\operatorname{Re} s > 1$,

$$\zeta(s)\Gamma(s) = \lim_{\epsilon \to 0} \int_{\epsilon}^{\infty} \frac{t^{s-1} \, dt}{e^t - 1}$$

$$= \lim_{\epsilon \to 0} (e^{2\pi i s} - 1)^{-1} \int_{L(\epsilon)} \frac{z^{s-1} \, dz}{e^z - 1}.$$

Note that the contour integral along $L(\epsilon)$ is the same provides that $0 < \epsilon < 2\pi$. Therefore we can drop the limit and it follows that for $\operatorname{Re} s > 1$,

$$\zeta(s)\Gamma(s) = (e^{2\pi i s} - 1)^{-1} \int_{L(\epsilon)} \frac{z^{s-1} \, dz}{e^z - 1}. \qquad \square$$

Proposition 2.2.3. *The Riemann zeta function* $\zeta(s)$ *has an analytic continuation given by*

$$\zeta(s) = e^{-\pi i s} \Gamma(1 - s) \frac{1}{2\pi i} \int_{L(\epsilon)} \frac{z^{s-1} \, dz}{e^z - 1}.$$

Proof. By the previous proposition, we have for $\operatorname{Re} s > 1$,

$$(e^{2\pi i s} - 1)\zeta(s)\Gamma(s)\Gamma(1 - s) = \Gamma(1 - s) \int_{L(\epsilon)} \frac{z^{s-1}\, dz}{e^z - 1}.$$

In light of the well-known formula

$$\Gamma(s)\Gamma(1 - s) = \frac{\pi}{\sin(\pi s)} = \frac{2\pi i}{e^{\pi i s} - e^{-\pi i s}} = \frac{2\pi i e^{\pi i s}}{e^{2\pi i s} - 1},$$

we have

$$\zeta(s) = e^{-\pi i s}\Gamma(1 - s)\frac{1}{2\pi i} \int_{L(\epsilon)} \frac{z^{s-1}\, dz}{e^z - 1}.$$

Note that the above contour integral is absolutely convergent for all $s \in \mathbb{C} \setminus \mathbb{Z}$. Thus it gives the analytic continuation of $\zeta(s)$ in the whole complex plane. $\qquad\square$

2.3. Bernoulli Numbers and Special Values of Riemann Zeta Function

Recall that the sequence of *Bernoulli numbers* B_n, $n = 0, 1, 2, 3, \ldots$, are defined by

$$\frac{t}{e^t - 1} = \sum_{n=0}^{\infty} \frac{B_n t^n}{n!}, \quad |t| < 2\pi.$$

They can be evaluated via the following recursion formula (see, for example, Rademacher [45, p. 6])

$$B_0 = 1; \quad \binom{n}{n-1}B_{n-1} + \binom{n}{n-2}B_{n-2} + \cdots + \binom{n}{0}B_0 = 0, \quad n \geq 2.$$

In particular, by a direct calculation, we get

$$B_0 = 1, \quad B_1 = -\frac{1}{2}, \quad B_2 = \frac{1}{6}, \quad B_3 = 0, \quad B_4 = -\frac{1}{30}.$$

Note that the function

$$F(t) = \frac{t}{e^t - 1} + \frac{t}{2}$$

is an even function, so $B_{2k+1} = 0$ for all positive integer k.

Bernoulli numbers appear naturally in the sums of the mth power of consecutive integers.

Proposition 2.3.1. *For a pair of positive integers m and N, we have*

$$S_m(N) = 1^m + 2^m + \cdots + (N-1)^m = \frac{1}{m+1} \sum_{\ell=0}^{m} \binom{m+1}{\ell} B_\ell N^{m+1-\ell}.$$

Proof. For every integer $k \geq 0$, we have

$$e^{kt} = 1 + kt + \frac{1}{2!}(kt)^2 + \cdots + \frac{1}{n!}(kt)^n + \cdots.$$

Summing together with k from 0 to $N-1$, it yields

$$\sum_{k=0}^{N-1} e^{kt} = N + S_1(N)t + \frac{1}{2!} S_2(N)t^2 + \cdots + \frac{1}{n!} S_n(N)t^n + \cdots.$$

It follows that $S_m(N)/m!$ is the coefficient of t^m in the asymptotic expansion at $t = 0$ of the function

$$F(t) = \sum_{k=0}^{N-1} e^{kt} = \frac{e^{Nt} - 1}{e^t - 1},$$

or it is equal to the coefficient of t^{m+1} in the Taylor expansion at $t = 0$ of the function

$$G(t) = tF(t) = \frac{te^{Nt}}{e^t - 1} - \frac{t}{e^t - 1}.$$

That is,

$$\frac{1}{m!} S_m(N) = \frac{1}{(m+1)!} \left\{ \sum_{\ell=0}^{m+1} \binom{m+1}{\ell} B_\ell N^{m+1-\ell} - B_{m+1} \right\}$$

$$= \frac{1}{(m+1)!} \sum_{\ell=0}^{m} \binom{m+1}{\ell} B_\ell N^{m+1-\ell}.$$

$\square$

Proposition 2.3.2. *For each nonnegative integer m, we have*

$$\zeta(-m) = (-1)^m \frac{B_{m+1}}{m+1}.$$

Proof. We begin with the analytic continuation of $\zeta(s)$ given by

$$\zeta(s) = e^{-\pi i s} \Gamma(1-s) \frac{1}{2\pi i} \int_{L(\epsilon)} \frac{z^{s-1}}{e^z - 1} dz.$$

When $s = -m$ is an integer, the integrations along $[\epsilon, \infty)$ twice in opposite directions cancel each other. So that

$$\zeta(-m) = (-1)^m m! \frac{1}{2\pi i} \int_{|z|=\epsilon} \frac{z^{-m-1}\, dz}{e^z - 1}$$

$$= (-1)^m m! \times \text{the coefficient of } z^m \text{ in the asymptotic expansion}$$

$$\text{at } z = 0 \text{ of the function } 1/(e^z - 1)$$

$$= (-1)^m m! \frac{B_{m+1}}{(m+1)!}$$

$$= (-1)^m \frac{B_{m+1}}{m+1}.$$

$\square$

Recall that the general procedure to obtain the special values of $\zeta(s)$ at nonpositive integers as follows:

Step 1. Consider the corresponding exponential series

$$\varphi(t) = \sum_{n=1}^{\infty} e^{-nt}.$$

The product $\zeta(s)\Gamma(s)$ is simply the Mellin transform of $\varphi(t)$, i.e.,

$$\zeta(s)\Gamma(s) = \int_0^{\infty} t^{s-1}\varphi(t)\, dt.$$

Step 2. The special value of $\zeta(s)$ at $s = -m$ is given by

$$\zeta(-m) = (-1)^m m! \times \text{the coefficient of } t^m \text{ in the}$$

$$\text{asymptotic expansion at } t = 0 \text{ of } \varphi(t).$$

Bernoulli polynomials $B_n(x)$, $n = 0, 1, 2, \ldots$, are defined by

$$B_n(x) = \sum_{k=0}^{n} \binom{n}{k} B_k x^{n-k},$$

or equivalently

$$\frac{te^{xt}}{e^t - 1} = \sum_{n=0}^{\infty} \frac{B_n(x)t^n}{n!}, \quad |t| < 2\pi.$$

Note that $B_n(x)$ is a polynomial of degree n with leading coefficient 1. The following relation is important for evaluating Bernoulli polynomials.

Proposition 2.3.3. *For each integer $n \geq 0$, we have*

$$\frac{d}{dx}B_{n+1}(x) = (n+1)B_n(x) \quad and \quad B_n(0) = B_n.$$

Proof. Differentiating both sides of the identity

$$\frac{te^{xt}}{e^t - 1} = \sum_{n=0}^{\infty} \frac{B_n(x)t^n}{n!}$$

with respect to x, we find that

$$\frac{t^2 e^{xt}}{e^t - 1} = \sum_{n=0}^{\infty} \frac{B_n'(x)t^n}{n!}.$$

It follows that for all integer $n \geq 0$,

$$\frac{B_n(x)}{n!} = \frac{B_{n+1}'(x)}{(n+1)!}$$

and hence

$$B_{n+1}'(x) = (n+1)B_n(x).$$

The second assertion follows from the definition of Bernoulli polynomials. $\square$

According to the above proposition, we deduce that

$$B_0(x) = 1,$$

$$B_1(x) = x - \frac{1}{2},$$

$$B_2(x) = x^2 - x + \frac{1}{6},$$

$$B_3(x) = x^3 - \frac{3x^2}{2} + \frac{x}{2}.$$

Proposition 2.3.4. *For each nonnegative integer n, we have*

$$B_n(1-x) = (-1)^n B_n(x) \quad and \quad B_{2n+1}\left(\frac{1}{2}\right) = 0.$$

Proof. The assertion follows from the following computations:

$$\sum_{n=0}^{\infty} \frac{B_n(1-x)t^n}{n!} = \frac{te^{(1-x)t}}{e^t - 1} = \frac{(-t)e^{x(-t)}}{e^{-t} - 1} = \sum_{n=0}^{\infty} \frac{B_n(x)(-t)^n}{n!}. \quad \square$$

Now if we consider *Hurwitz zeta function* $\zeta(s; \delta)$ defined by

$$\zeta(s; \delta) = \sum_{n=0}^{\infty} (n + \delta)^{-s}, \quad \mathrm{Re}\, s > 1, \quad \delta > 0,$$

we obtain among other things that the special values of $\zeta(s; \delta)$ at non-positive integers are given by

$$\zeta(1 - m; \delta) = -\frac{B_m(\delta)}{m}$$

for any positive integer m. The details of the proof will be left as an exercise.

Proposition 2.3.5. *For any positive integer k, we have*

$$B_m(kx) = k^{m-1} \sum_{j=0}^{k-1} B_m\left(\frac{j}{k} + x\right).$$

Proof. For the time being, we assume that $x > 0$. Then for $\mathrm{Re}\, s > 1$,

$$\zeta(s; kx) = \sum_{n=0}^{\infty} (n + kx)^{-s}$$

$$= \sum_{j=0}^{k-1} \sum_{m=0}^{\infty} (km + j + kx)^{-s}$$

$$= k^{-s} \sum_{j=0}^{k-1} \zeta\left(s; \frac{j}{k} + x\right).$$

All the Hurwitz zeta functions above have analytic continuations (see, for example, Apostol [3, p. 251]). Setting $s = 1 - m$, we find that

$$-\frac{B_m(kx)}{m} = -k^{m-1} \sum_{j=0}^{k-1} \frac{B_m(j/k + x)}{m}. \tag{2.3.1}$$

Thus our assertion holds for $x > 0$. However, since both sides of the equality are polynomials, (2.3.1) must be true for all real numbers x. $\square$

Setting $x = 0$ in (2.3.1), we deduce

Corollary 2.3.6. *For any positive integer k, we have*

$$(1 - k^{m-1})B_m = \sum_{j=1}^{k-1} B_m\left(\frac{j}{k}\right).$$

For any positive integer n, the Bernoulli polynomial $B_{2n}(\{x\})$ has a Fourier series expansion.

Proposition 2.3.7. *Let $\{x\} = x - [x]$ be the fraction part of x. Then for any positive integer n, we have*

$$\sum_{k=1}^{\infty} \frac{\cos(2k\pi x)}{k^{2n}} = \frac{(-1)^{n-1}(2\pi)^{2n} B_{2n}(\{x\})}{2(2n)!}.$$

Proof. The function $B_{2n}(\{x\})$ is a periodic function with period 1. So it has a Fourier expansion

$$B_{2n}(\{x\}) = \frac{1}{2}a_0 + \sum_{k=1}^{\infty} a_k \cos(2k\pi x) + \sum_{k=1}^{\infty} b_k \sin(2k\pi x),$$

where

$$a_k = 2 \int_0^1 B_{2n}(x) \cos(2k\pi x)\, dx$$

and

$$b_k = 2 \int_0^1 B_{2n}(x) \sin(2k\pi x)\, dx.$$

When $k = 0$, we have

$$a_0 = 2 \int_0^1 B_{2n}(x)\, dx = \frac{2}{2n+1} \left(B_{2n+1}(1) - B_{2n+1}(0) \right) = 0.$$

For $k \geq 1$ and $n \geq 2$, with integration by parts, we get

$$a_k = \frac{2}{-2k\pi} B_{2n}(x) \sin(2k\pi x) \Big|_0^1 + \frac{2n}{2k\pi} 2 \int_0^1 B_{2n-1}(x) \sin(2k\pi x)\, dx$$

$$= 0 + \frac{(-2n)(2n-1)}{(2k\pi)^2} 2 \int_0^1 B_{2n-2}(x) \cos(2k\pi x)\, dx.$$

An induction on n leads to the conclusion

$$a_k = \frac{(-1)^{n-1} 2(2n)!}{(2k\pi)^{2n}}.$$

On the other hand, we also have for $n \geq 2$,

$$b_k = \frac{-2n(2n-1)}{(2k\pi)^2} 2 \int_0^1 B_{2n-2}(x) \sin(2k\pi x)\, dx.$$

An induction on n leads to the conclusion $b_k = 0$. $\square$

2.4. Functional Equation of $\zeta(s)$

E. Hecke was probably one of the first few mathematicians to realize that there is a transformation formula of a modular form behind a functional equation of a zeta function. There is no exception for the Riemann zeta function. In order to derive the transformation formula of the theta series corresponding to Riemann zeta function, we need the following useful classical formula.

Theorem 2.4.1 (Poisson Summation Formula). *For any $f(x)$ in $\mathscr{S}(\mathbb{R}^n)$, the space of rapidly decreasing function on $\mathbb{R}^n$, we have*

$$\sum_{\lambda \in \mathbb{Z}^n} f(\lambda) = \sum_{\mu \in \mathbb{Z}^n} \widehat{f}(\mu),$$

where $\widehat{f}$ is the Fourier transform of f defined by

$$\widehat{f}(y) = \int_{\mathbb{R}^n} f(x) e^{2\pi i \langle x, y \rangle} \, dx.$$

Here $\langle x, y \rangle$ is the usual inner product of $\mathbb{R}^n$.

Proof. Consider the function given by

$$g(x) = \sum_{\lambda \in \mathbb{Z}^n} f(\lambda + x).$$

We have

$$g(x + \mu) = g(x)$$

for all $\mu \in \mathbb{Z}^n$. Thus $g(x)$ is a periodic function in x and hence it has a Fourier expansion

$$g(x) = \sum_{\mu \in \mathbb{Z}^n} c_\mu e^{-2\pi i \langle x, \mu \rangle}.$$

The Fourier coefficient c_μ is

$$c_\mu = \int_{\mathbb{R}^n / \mathbb{Z}^n} g(x) e^{2\pi i \langle x, \mu \rangle} \, dx$$

$$= \int_{\mathbb{R}^n / \mathbb{Z}^n} \sum_{\lambda \in \mathbb{Z}^n} f(\lambda + x) e^{2\pi i \langle x, \mu \rangle} \, dx$$

$$= \int_{\mathbb{R}^n} f(x) e^{2\pi i \langle x, \mu \rangle} \, dx$$

$$= \widehat{f}(\mu).$$

Consequently, we have

$$\sum_{\lambda \in \mathbb{Z}^n} f(\lambda + x) = \sum_{\mu \in \mathbb{Z}^n} \widehat{f}(\mu) e^{-2\pi i \langle x, \mu \rangle}.$$

Setting $x = 0$, we get our assertion.   □

Proposition 2.4.2. *For any positive number t, one has*

$$\sum_{n \in \mathbb{Z}} e^{-\pi n^2 / t} = \sqrt{t} \sum_{n \in \mathbb{Z}} e^{-\pi n^2 t}.$$

Proof. It follows from the previous proposition and the Fourier transform of $e^{-\pi x^2 / t}$ is $\sqrt{t} e^{-\pi y^2 t}$.   □

Proposition 2.4.3 (The Functional Equation of the Riemann Zeta Function). *For $\operatorname{Re} s > 1$, let*

$$X(s) = \pi^{-s/2} \Gamma\left(\frac{s}{2}\right) \zeta(s).$$

Then $X(s)$ has its analytic continuation in the whole complex plane. Furthermore, we have

$$X(1 - s) = X(s).$$

Proof. For $t > 0$, let

$$g(t) = \sum_{n=1}^{\infty} e^{-\pi n^2 t}.$$

Then for $\operatorname{Re} s > 1$, we have

$$
\begin{aligned}
X(s) &= \int_0^{\infty} t^{s/2-1} g(t)\, dt \\
&= \int_1^{\infty} t^{s/2-1} g(t)\, dt + \int_0^1 t^{s/2-1} g(t)\, dt \\
&= \int_1^{\infty} t^{s/2-1} g(t)\, dt + \int_1^{\infty} t^{-s/2-1} g\left(\frac{1}{t}\right) dt.
\end{aligned}
$$

By the previous proposition, we have

$$2g\left(\frac{1}{t}\right) + 1 = \sqrt{t}\, [2g(t) + 1].$$

It is equivalent to

$$g\left(\frac{1}{t}\right) = \sqrt{t} g(t) + \frac{1}{2}\sqrt{t} - \frac{1}{2}.$$

Consequently, we have for $\operatorname{Re} s > 1$,

$$X(s) = \int_1^\infty t^{s/2-1} g(t)\, dt + \int_1^\infty t^{-(s+1)/2} g(t)\, dt - \frac{1}{1-s} - \frac{1}{s}.$$

Both improper integrals are absolutely convergent for all $s \in \mathbb{C}$. The above expression for $X(s)$ give the analytic continuation of $X(s)$ in the whole complex plane. Furthermore, we have

$$X(1-s) = X(s). \qquad \qquad \square$$

Corollary 2.4.4. *For any positive integer m, we have*

$$\zeta(2m) = \frac{(-1)^{m-1} (2\pi)^{2m} B_{2m}}{2(2m)!}. \qquad (2.4.1)$$

Proof. It follows from Proposition 2.3.2 that

$$\zeta(1 - 2m) = -\frac{B_{2m}}{2m}.$$

Using the functional equation of $\zeta(s)$, we obtain (2.4.1). $\qquad \square$

For another proof of (2.4.1), see Remark 1.3.5.

2.5. Exercises

1. Let γ be the Euler constant defined by

$$\gamma = \lim_{n\to\infty} \left(\sum_{k=1}^n \frac{1}{k} - \log n \right).$$

   Prove that

$$\lim_{n\to\infty} \left(1 + \frac{1}{3} + \frac{1}{5} + \cdots + \frac{1}{2n-1} - \frac{1}{2} \log n \right) = \frac{\gamma}{2} + \log 2.$$

2. Based on the classical Kronecker limit formula

$$\lim_{s\to 1^+} \left(\zeta(s) - \frac{1}{s-1} \right) = \gamma,$$

   prove that for $\delta > 0$,

$$\lim_{s\to 1^+} \left(\zeta(s; \delta) - \frac{1}{s-1} \right) = -\frac{\Gamma'(\delta)}{\Gamma(\delta)}.$$

3. Define

$$\Lambda(n) = \begin{cases} \log p, & \text{if } n = p^m \text{ for some prime } p \text{ and positive integer } m; \\ 0, & \text{otherwise} \end{cases}$$

and

$$\psi(x) = \sum_{1 \le n \le x} \Lambda(n) = \sum_{p \le x} \left[\frac{\log x}{\log p} \right] \log p.$$

Prove that if $\psi(x) \sim x$ as $x \to \infty$, then $\pi(x) \sim x/\log x$ as $x \to \infty$.

4. Prove that for $\text{Re } s > 1$,

$$\frac{\zeta'(s)}{\zeta(s)} = -\sum_{m=1}^{\infty} \sum_{p} (\log p) p^{-ms} = -\sum_{n=1}^{\infty} \frac{\Lambda(n)}{n^s}.$$

5. Find the special values at negative integers of the zeta function

$$Z(s) = \sum_{n_1=1}^{\infty} \sum_{n_2=1}^{\infty} (n_1 + n_2)^{-s}, \quad \text{Re } s > 2.$$

6. Prove that the special values at nonnegative integers $s = -m$, $m = 1, 2, 3, \ldots$, of Hurwitz zeta function are given by

$$\zeta(1 - m; \delta) = -\frac{B_m(\delta)}{m}.$$

Chapter 3

Zeta Functions of Modular Forms

3.1. Theta Series as Examples of Modular Forms

We have seen that Eisenstein series are modular forms. Another source of modular forms comes from theta series which are sums of exponential functions over certain lattices. Let S be an $m \times m$ positive definite integral symmetric matrix. Set

$$S[g] = {}^t g S g, \quad g \in \mathbb{R}^n.$$

Here we have the following equivalent conditions for S to be positively definite (see, for example, Andrianov [2] or Freitag [31]).

(a) There exists $\epsilon > 0$ such that

$$S[g] \geq \epsilon |g|^2$$

for all $g \in \mathbb{R}^m$, $g \neq 0$.

(b) All the eigenvalues of S are positive real numbers.

(c) Let $S^{(i)}$ be the $i \times i$ submatrix in the left-upper block of S. Then

$$\det S^{(i)} > 0$$

for all $i = 1, 2, \ldots, n$.

The *theta series* $\vartheta(S, z)$ defined by

$$\vartheta(S, z) = \sum_{g \in \mathbb{Z}^m} e^{\pi i S[g] z}, \quad z \in \mathcal{H}$$

converges uniformly on any compact subset of the upper half-plane $\mathcal{H}$ and it defines a holomorphic function there. Also

$$\vartheta(S, z) = \sum_{n=0}^{\infty} A(n) e^{2\pi i n z},$$

where $A(n)$ is the number of integral solutions to the quadratic equation

$$S[g] = 2n.$$

Here we need further conditions imposed on S so that $\vartheta(S, z)$ is a modular form of weight $m/2$ when $m \equiv 0 \pmod{8}$.

Proposition 3.1.1. *The theta series $\vartheta(S, z)$ is a modular form of weight $m/2$ if S satisfies the following further conditions:*

(a) *$S[g]$ is even for all $g \in \mathbb{Z}^m$;*

(b) *S is unimodular, that is, $\det S = 1$.*

Proof. With the first condition, we have

$$\vartheta(S, z + 1) = \sum_{g \in \mathbb{Z}^m} e^{\pi i S[g]} e^{\pi i S[g] z} = \sum_{g \in \mathbb{Z}^m} e^{\pi i S[g] z} = \vartheta(S, z).$$

Note that the conditions (a) and (b) imply that (see Freitag [31])

$$m \equiv 0 \pmod{8}.$$

Now we want to prove that

$$\vartheta(S, iy^{-1}) = y^{m/2} \vartheta(S, iy).$$

The Fourier transform of the function $e^{-\pi S[x] y^{-1}}$ is given by (see Freitag [31])

$$(\det S)^{-1/2} y^{m/2} e^{-\pi S^{-1}[w] y}$$

by a direct calculation. It follows from Poisson summation formula that

$$\vartheta(S, iy^{-1}) = y^{m/2} \sum_{g \in \mathbb{Z}^m} e^{-\pi S^{-1}[g]y}$$

$$= y^{m/2} \sum_{g \in \mathbb{Z}^m} e^{-\pi S^{-1}[Sg]y}$$

$$= y^{m/2} \sum_{g \in \mathbb{Z}^m} e^{-\pi S[g]y}$$

$$= y^{m/2} \vartheta(S, iy).$$

Consequently, we have

$$\vartheta\left(S, -\frac{1}{z}\right) = z^{m/2} \vartheta(S, z). \qquad \square$$

Remark 3.1.2. Two positive definite symmetric matrices S_1 and S_2 are equivalent if there exists $U \in \mathrm{SL}_m(\mathbb{Z})$ such that

$$S_1[U] = S_2.$$

If S_1 and S_2 are equivalent, then

$$\vartheta(S_1, z) = \vartheta(S_2, z)$$

since the lattice $\mathbb{Z}^m$ is invariant under the left multiplication of $U \in \mathrm{SL}_m(\mathbb{Z})$.

A well-known 8×8 integral symmetric matrix, which is positive definite, even and unimodular, is given by

$$S = \begin{bmatrix} 2E_4 & B \\ {}^tB & 2E_4 \end{bmatrix} \quad \text{with} \quad B = \begin{bmatrix} 0 & -1 & -1 & -1 \\ 1 & -1 & 1 & 0 \\ 1 & 0 & -1 & 1 \\ 1 & 1 & 0 & -1 \end{bmatrix}$$

and

$$S^{-1} = \begin{bmatrix} 2 & 0 & 0 & 0 & 0 & 1 & 1 & 1 \\ 0 & 2 & 0 & 0 & -1 & 1 & -1 & 0 \\ 0 & 0 & 2 & 0 & -1 & 0 & 1 & -1 \\ 0 & 0 & 0 & 2 & -1 & -1 & 0 & 1 \\ 0 & -1 & -1 & -1 & 2 & 0 & 0 & 0 \\ 1 & 1 & 0 & -1 & 0 & 2 & 0 & 0 \\ 1 & -1 & 1 & 0 & 0 & 0 & 2 & 0 \\ 1 & 0 & -1 & 1 & 0 & 0 & 0 & 2 \end{bmatrix}.$$

Consequently, $\vartheta(S,z)$ is a modular form of weight 4. On the other hand, the normalized Eisenstein series

$$E_4(z) = 1 + 240 \sum_{n=1}^{\infty} \left(\sum_{d|n} d^3 \right) e^{2\pi i n z}$$

is also a modular form of weight 4. As we shall prove later that the vector space of modular forms of weight 4 has dimension 1. Hence,

$$\vartheta(S,z) = E_4(z).$$

So we have the following interesting consequence.

Proposition 3.1.3. *Let $A(S, 2n)$ be the number of integral solutions to the equation*

$$S[g] = 2n, \quad g \in \mathbb{Z}^m.$$

Then

$$A(S, 2n) = 240 \sum_{d|n} d^3.$$

3.2. Zeta Functions Attached to Modular Forms

Let

$$f(z) = \sum_{n=0}^{\infty} a_n e^{2\pi i n z}$$

be a modular form of weight k with respect to the full modular group $\Gamma = \mathrm{SL}_2(\mathbb{Z})$. The zeta function attached to f is given by

$$Z(f, s) = \sum_{n=1}^{\infty} a_n n^{-s}.$$

As we shall see in Proposition 3.2.2, $Z(f, s)$ is absolutely convergent when $\mathrm{Re}\, s > k$ since

$$a_n = O(n^{k-1}).$$

The function $Z(f, s)$ is also related to $f(z)$ via the transform

$$(2\pi)^{-s} Z(f, s) \Gamma(s) = \int_0^{\infty} y^{s-1} \{f(iy) - a_0\}\, dy.$$

Hence, it has an analytic continuation in the whole complex plane and satisfies a functional equation like the classical Riemann zeta function.

Proposition 3.2.1. *If the function*

$$f(z) = \sum_{n=1}^{\infty} a_n e^{2\pi i n z}$$

is a cusp form of weight k, then

$$a_n = O(n^{k/2}).$$

Proof. We have

$$|f(z)| = |e^{2\pi i z}| \left| \sum_{n=1}^{\infty} a_n e^{2\pi i (n-1) z} \right| = O(e^{-2\pi y}).$$

Let

$$\varphi(z) = |f(z)| y^{k/2}.$$

Then for all $\left[\begin{smallmatrix} a & b \\ c & d \end{smallmatrix} \right] \in \mathrm{SL}_2(\mathbb{Z})$, we have

$$\varphi\left(\frac{az+b}{cz+d} \right) = \varphi(z).$$

In addition, φ is continuous on the fundamental domain

$$\mathscr{F} = \left\{ z \in \mathscr{H} \,\middle|\, |z| \geq 1, -\frac{1}{2} \leq \mathrm{Re}\, z \leq \frac{1}{2} \right\}$$

and $\varphi(z) \to 0$ as $y \to +\infty$. This implies that φ must be bounded on the whole upper half-plane, so there exists $M > 0$ such that

$$|f(z)| \leq M y^{-k/2}$$

for all $z \in \mathscr{H}$. Now

$$\int_0^1 f(z) e^{-2\pi i n x} \, dx = a_n e^{-2\pi n y}.$$

Thus

$$|a_n| e^{-2\pi n y} \leq \int_0^1 |f(z)| \, dx \leq M y^{-k/2}$$

or it is equivalent to

$$|a_n| \leq M y^{-k/2} e^{2\pi n y}.$$

Setting $y = 1/n$, we get

$$|a_n| \leq M' n^{k/2}. \qquad \square$$

Proposition 3.2.2. *If*

$$f(z) = \sum_{n=0}^{\infty} a_n e^{2\pi i n z}$$

is a modular form of weight k with respect to $\mathrm{SL}_2(\mathbb{Z})$, *then*

$$a_n = O(n^{k-1}).$$

Proof. We can express $f(z)$ as

$$a_0 E_k(z) + \varphi(z),$$

where $\varphi(z)$ is a cusp form of the same weight. Note that

$$E_k(z) = 1 - \frac{2k}{B_k} \sum_{n=1}^{\infty} \sigma_{k-1}(n) e^{2\pi i n z}.$$

The Fourier coefficients of $E_k(z)$ are constant multiples of $\sigma_{k-1}(n)$. But

$$n^{k-1} \leq \sigma_{k-1}(n) = n^{k-1} \sum_{d|n} \frac{1}{(n/d)^{k-1}} \leq n^{k-1} \zeta(k-1).$$

It follows that

$$a_n = O(n^{k-1}). \qquad \square$$

Proposition 3.2.3. *The zeta function $Z(f, s)$ attached to a modular form*

$$f(z) = \sum_{n=0}^{\infty} a_n e^{2\pi i n z}$$

of weight k is absolutely convergent for $\operatorname{Re} s > k$ *and it has analytic continuation which has possible simple poles at $s = k$ and $s = 0$. Furthermore, it satisfies the functional equation*

$$(2\pi)^{-s} \Gamma(s) Z(f, s) = i^k (2\pi)^{k-s} \Gamma(k-s) Z(f, k-s).$$

Proof. For $\operatorname{Re} s > k$, one has

$$(2\pi)^{-s} \Gamma(s) Z(f, s) = \int_0^{\infty} y^{s-1} \{f(iy) - a_0\} \, dy$$

$$= \int_1^{\infty} y^{s-1} \{f(iy) - a_0\} \, dy + \int_0^1 y^{s-1} \{f(iy) - a_0\} \, dy$$

$$= \int_1^{\infty} y^{s-1} \{f(iy) - a_0\} \, dy - \frac{a_0}{s} + \int_0^1 y^{s-1} f(iy) \, dy$$

$$= \int_1^{\infty} y^{s-1} \{f(iy) - a_0\} \, dy - \frac{a_0}{s} + \int_1^{\infty} y^{-s-1} f(iy^{-1}) \, dy.$$

Since f is a modular form of weight k, we deduce that

$$f(iy^{-1}) = (iy)^k f(iy).$$

If follows that for $\operatorname{Re} s > k$,

$$\int_1^\infty y^{-s-1} f(iy^{-1})\, dy = i^k \int_1^\infty y^{k-s-1} f(iy)\, dy$$

$$= i^k \int_1^\infty y^{k-s-1} \{f(iy) - a_0\}\, dy - \frac{i^k a_0}{k-s}.$$

Therefore for $\operatorname{Re} s > k$, we have

$$(2\pi)^{-s}\Gamma(s)Z(f,s) = \int_1^\infty y^{s-1} \{f(iy) - a_0\}\, dy$$

$$+ i^k \int_1^\infty y^{k-s-1} \{f(iy) - a_0\}\, dy - \frac{a_0}{s} - \frac{i^k a_0}{k-s}.$$

The improper integrals in the above expression converge for any $s \in \mathbb{C}$. So the above expression gives the analytic continuation of $(2\pi)^{-s}\Gamma(s)Z(f,s)$ for all s in the complex plane. Furthermore, we have

$$(2\pi)^{-(k-s)}\Gamma(k-s)Z(f,k-s) = i^k(2\pi)^{-s}\Gamma(s)Z(f,s).$$

Our assertion then follows since k is an even integer. $\qquad\square$

The zeta function attached to the normalized Eisenstein series $E_k(z)$, up to a constant factor, is

$$\sum_{n=1}^\infty \left(\sum_{d|n} d^{k-1}\right) n^{-s} = \zeta(s)\zeta(s-k+1).$$

Note that $\zeta(s)\zeta(s-k+1)$ has the Euler product

$$\zeta(s)\zeta(s-k+1) = \prod_p \left[\left(1-p^{-s}\right)\left(1-p^{-s+k-1}\right)\right]^{-1}.$$

So another problem arises: when will a zeta function attached to a modular form admits an Euler product? The answer is not so obvious. It turns out that the modular form must be a common eigenfunction of Hecke operators on modular forms. We will discuss these topics in detail in the next section.

3.3. Hecke Operators on Modular Forms

Hecke operators are the most important operators on modular forms, which map a modular form into a finite sum of modular forms. Here we begin with the definition from an algebraic point of view.

Given any positive integer n, let $\Gamma = \mathrm{SL}_2(\mathbb{Z})$ and

$$M = \begin{bmatrix} n & 0 \\ 0 & 1 \end{bmatrix}.$$

Then Γ contains its subset $\Gamma \cap M^{-1}\Gamma M$ as a subgroup of finite index (see Shimula [48]). Suppose that Γ has the coset decomposition

$$\Gamma = \bigcup_{j=1}^{p} (\Gamma \cap M^{-1}\Gamma M)\delta_j.$$

Then it follows that

$$M^{-1}\Gamma M\Gamma = \bigcup_{j=1}^{p} M^{-1}\Gamma M\delta_j$$

and hence

$$\Gamma M\Gamma = \bigcup_{j=1}^{p} \Gamma M_j \quad \text{with} \quad M_j = M\delta_j$$

gives the coset decomposition of the double coset $\Gamma M\Gamma$.

If $f(z)$ is a modular form of weight k with respect to the full modular group $\Gamma = \mathrm{SL}_2(\mathbb{Z})$, we define the *Hecke operator* $T(n)$ by

$$T(n)(f(z)) = n^{k-1} \sum_{j=1}^{p} j(M_j, z)^{-k} f(M_j(z)).$$

The coset representatives may be chosen as (see Serre [47])

$$\begin{bmatrix} a & b \\ 0 & d \end{bmatrix}, \quad ad = n, \quad 0 \le b < d.$$

Therefore, we also have

$$T(n)(f(z)) = n^{k-1} \sum_{ad=n} \sum_{b=0}^{d-1} d^{k-1} f\left(\frac{az+b}{d}\right).$$

Remark 3.3.1. The double coset $\Gamma M \Gamma$ is nothing but the set of 2×2 integral matrices of determinant n. With elementary row operations, we are able to reduce any matrix in $\Gamma M \Gamma$ to a triangular form as shown above.

Proposition 3.3.2. *If f is a modular form of weight k with respect to the full modular group $\Gamma = \mathrm{SL}_2(\mathbb{Z})$, then $T(n)(f)$ is also a modular form of weight k.*

Proof. If $\{M_1, M_2, \ldots, M_p\}$ is a set of coset representatives in the decomposition of $\Gamma M \Gamma$, then $\{M_1 K, M_2 K, \ldots, M_p K\}$ is also a set of coset representatives for any K in Γ. Then it follows that

$$T(n)(f(K(z))) = n^{k-1} \sum_{j=1}^{p} j(M_j, K(z))^{-k} f(M_j K(z))$$

$$= j(K, z)^k n^{k-1} \sum_{j=1}^{p} j(M_j K, z)^{-k} f(M_j K(z))$$

$$= j(K, z)^k T(n)(f(z)).$$

Next, we prove that $T(n)(f)$ has a Fourier expansion. If f has a Fourier expansion

$$\sum_{m=0}^{\infty} a(m) e^{2\pi i m z},$$

then

$$T(n)(f(z)) = n^{k-1} \sum_{ad=n} \sum_{b=0}^{d-1} \sum_{m=0}^{\infty} a(m) d^{-k} e^{2\pi i m (az+b)/d}.$$

Note that the inner sum is

$$\sum_{b=0}^{d-1} e^{2\pi i m b/d} = \begin{cases} d, & \text{if } d \mid m; \\ 0, & \text{otherwise.} \end{cases}$$

It follows that

$$T(n)(f(z)) = \sum_{d \mid n} \sum_{d \mid m} \sum_{m=0}^{\infty} a(m) \left(\frac{n}{d}\right)^{k-1} e^{2\pi i m n z/d^2}$$

$$= \sum_{\ell=0}^{\infty} \sum_{d \mid \ell} \sum_{d \mid n} d^{k-1} a\left(\frac{\ell n}{d^2}\right) e^{2\pi i \ell z}.$$

This proves that $T(n)(f)$ also has a Fourier expansion and hence $T(n)(f)$ is indeed a modular form of weight k. $\square$

Here we shall describe Hecke operators from another point of view. Let $\mathscr{R}$ be the set of lattices of rank 2 in $\mathbb{C}$. For any lattice $\mathscr{L}$ in $\mathscr{R}$ and a positive integer n, we define

$$T(n)(\mathscr{L}) = \sum_{[\mathscr{L}:\mathscr{L}']=n} \mathscr{L}'.$$

The Hecke operator $T(n)$ transforms a lattice $\mathscr{L}$ into a sum of its sublattices of index n. Hence it defines a mapping from the free abelian group with $\mathscr{R}$ as a generator into itself. The sum on the right hand side is finite since all $\mathscr{L}'$ contains $n\mathscr{L}$ and the number is equal to the number of subgroups of order n in $\mathscr{L}/n\mathscr{L} \cong (\mathbb{Z}/n\mathbb{Z})^2$.

Also we define another type of mappings of $\mathscr{R}$ into itself by

$$R_\lambda(\mathscr{L}) = \lambda\mathscr{L}, \quad \lambda \in \mathbb{C} \cup \{\infty\}, \quad \mathscr{L} \in \mathscr{R},$$

where R_λ is called the *homothety operator*. Then we have the following basic properties for $T(n)$ and R_λ.

(1) $R_\lambda R_\mu = R_{\lambda\mu}$, $\lambda,\, \mu \in \mathbb{C} \cup \{\infty\}$.

(2) $R_\lambda T(n) = T(n)R_\lambda$, $\lambda \in \mathbb{C} \cup \{\infty\}$, $n \in \mathbb{Z}^+$.

(3) $T(m)T(n) = T(mn)$ if m and n are relatively prime.

(4) For any prime number p and positive integer n,

$$T(p^n)T(p) = T(p^{n+1}) + pT(p^{n-1})R_p.$$

For the details of proofs of above, see [47].

Recall that the Eisenstein series $G_k(z)$ is a function on a lattice of $\mathbb{C}$ given by

$$G_k(z) = \sum_{\lambda \in \mathscr{L} \backslash \{0\}} \lambda^{-k}$$

with

$$\mathscr{L} = \{mz + n \mid m, n \in \mathbb{Z}\}.$$

Let p be a prime number. The Hecke operator $T(p)$ on $G_k(z)$ is given by

$$T(p)(G_k(z)) = p^{k-1} \sum_{[\mathscr{L}:\mathscr{L}']=p} \sum_{\lambda \in \mathscr{L}' \setminus \{0\}} \lambda^{-k}.$$

Suppose $\lambda \in \mathscr{L} \setminus \{0\}$ and $\lambda \in p\mathscr{L}$. Then λ belongs to each of $p+1$ sublattices of $\mathscr{L}$ of index p, its contribution in $T(p)(G_k(z))$ is

$$(p+1)\lambda^{-k}.$$

If $\lambda \in \mathscr{L} \setminus p\mathscr{L}$, then λ belongs to one of sublattice of index p and its contribution is λ^{-k}. Therefore, we have

$$\begin{aligned}
T(p)(G_k(z)) &= p^{k-1}\left(G_k(z) + p \sum_{\lambda \in p\mathscr{L} \setminus \{0\}} \lambda^{-k}\right) \\
&= p^{k-1}\left(1 + p^{-k+1}\right) G_k(z) \\
&= \left(p^{k-1} + 1\right) G_k(z).
\end{aligned}$$

Thus $G_k(z)$ is a common eigenfunction of Hecke operators $T(p)$ for all prime number p.

With the help from the consideration of Hecke operators on lattices, we obtain the following further properties of Hecke operators.

(1) $T(m)T(n)(f) = T(mn)(f)$ if $(m, n) = 1$.

(2) $T(p)T(p^n)(f) = T(p^{n+1})(f) + p^{k-1}T(p^{n-1})(f)$ if p is a prime and n is a positive integer.

Now suppose that

$$f(z) = \sum_{m=0}^{\infty} a(m)e^{2\pi imz}$$

is a modular form of weight k and it is a common eigenfunction of Hecke operator $T(n)$ with

$$T(n)(f(z)) = \lambda(n)f(z)$$

and $a(1) = 1$. Then the Fourier coefficient of $e^{2\pi iz}$ in $T(n)(f(z))$ is

$$\sum_{d|(n,1)} d^{k-1}a\left(\frac{n}{d^2}\right) = a(n).$$

On the other hand, it is equal to

$$\lambda(n)a(1) = \lambda(n).$$

Thus, we find that

$$\lambda(n) = a(n).$$

By the first property of Hecke operators shown above, we have

$$a(mn) = a(m)a(n)$$

if m and n are relatively prime. Hence it follows that

$$\sum_{n=1}^{\infty} a(n)n^{-s} = \prod_{p} \left(1 + a(p)p^{-s} + a(p^2)p^{-2s} + \cdots + a(p^k)p^{-ks} + \cdots \right).$$

Also the second condition implies that

$$a(p)a(p^n) = a(p^{n+1}) + p^{k-1}a(p^{n-1})$$

and hence

$$(1 + a(p)p^{-s} + a(p^2)p^{-2s} + \cdots + a(p^k)p^{-ks} + \cdots)(1 - a(p)p^{-s} + p^{k-1-2s}) = 1.$$

Consequently, the zeta function attached to the modular form f has an Euler product

$$\sum_{n=1}^{\infty} a(n)n^{-s} = \prod_{p} \left(1 - a(p)p^{-s} + p^{k-1-2s}\right)^{-1}.$$

This is precisely asserted by the well-known *Hecke correspondence* on modular forms: The zeta function attached to a modular form satisfies a functional equation and it has an Euler product if and only if it is a common eigenfunction of Hecke operators.

3.4. Exercises

1. The well-known Fibonacci numbers $\{F_n\}_{n=0}^{\infty}$ are defined recursively by

$$F_0 = 1, \quad F_1 = 1, \quad F_n = F_{n-1} + F_{n-2} \quad \text{for } n \geq 2.$$

Consider the generating function

$$F(x) = \sum_{n=0}^{\infty} F_n x^n$$

associated with $\{F_n\}_{n=0}^{\infty}$ and prove that

$$F(x) = x^2 F(x) + x F(x) + x,$$

and give the explicit formula of F_n.

2. Let p be a prime number and n a positive integer. Prove that

$$\sum_{\substack{1 \le v < p^n \\ (v,p)=1}} e^{2\pi i v/p^n} = \begin{cases} -1, & \text{if } n = 1; \\ 0, & \text{if } n \ge 2. \end{cases}$$

3. Let S be an $m \times m$ positive definite integral symmetric matrix. Show that the Fourier transform of the function $f(x) = e^{-\pi S[x]y^{-1}}$, $x \in \mathbb{R}^m$, $y > 0$ is given by

$$\widehat{f}(w) = (\det S)^{-1/2} y^{m/2} e^{-\pi S^{-1}[w]y}.$$

4. Let S be the 8×8 integral symmetric matrix as given in page 35 and $T = \begin{bmatrix} S & 0 \\ 0 & S \end{bmatrix}$. Show that $\vartheta(T, z) = E_8(z) = [E_4(z)]^2$ and then find $A(T, 2n)$ for any positive integer n.

5. Let S be an $m \times m$ positive definite integral symmetric matrix of determinant 1. If S is even, prove that $m \equiv 0 \pmod{8}$ from the formula

$$\vartheta\left(S, \frac{z-1}{z}\right) = \vartheta\left(S, -\frac{1}{z}\right) = \left(\sqrt{\frac{z}{i}}\right)^m \vartheta(S, z).$$

Chapter 4

Dimension Formulæ

4.1. Dimension Formulæ for Modular Forms

Suppose that f is a holomorphic function defined on $\mathcal{H}$. At each point $z_0 \in \mathcal{H}$, f has a power series expansion

$$f(z) = \sum_{n=0}^{\infty} b_n (z - z_0)^n, \quad |z - z_0| < \delta.$$

Now we define $v_{z_0}(f)$ as the least integer n such that $b_n \neq 0$.

On the other hand, if f also has a Fourier expansion of the form

$$f(z) = \sum_{n=0}^{\infty} a_n e^{2\pi i n z},$$

we define $v_\infty(f)$ as the least integer n such that $a_n \neq 0$.

To find the total number of zeros of a holomorphic function enclosed by a simple closed curve in counterclockwise direction, we simply integrate

$$\frac{1}{2\pi i} \frac{df}{f}$$

along the curve. Indeed, we have from the residue theorem that

$$\frac{1}{2\pi i} \int_C \frac{df}{f} = \sum_p v_p(f),$$

where p ranges over all zeros of f inside the curve C.

Here we apply the integration to modular forms.

Proposition 4.1.1. *Let f be a nonzero modular form of weight k. Then*

$$v_\infty(f) + \frac{1}{2}v_i(f) + \frac{1}{3}v_\rho(f) + \sum_{p \in \mathscr{F}} v_p(f) = \frac{k}{12},$$

where $\mathscr{F}$ is the fundamental domain given by

$$\mathscr{F} = \left\{ z \in \mathscr{H} \ \middle| \ |z| \geq 1 \ \text{and} \ -\frac{1}{2} \leq \operatorname{Re} z \leq \frac{1}{2} \right\}.$$

Proof. We will integrate $f'/(2\pi i f)$ along the boundary of a truncated domain of $\mathscr{F}$ as shown below.

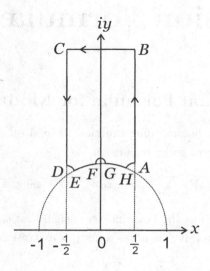

Figure 4.1: Boundary of a truncated domain of $\mathscr{H}$.

As $f(z+1) = f(z)$, we have

$$\frac{1}{2\pi i} \int_{\overline{AB}} \frac{df}{f} + \frac{1}{2\pi i} \int_{\overline{CD}} \frac{df}{f} = 0.$$

Also

$$\frac{1}{2\pi i} \int_{\overline{BC}} \frac{df}{f} = -v_\infty(f), \quad \frac{1}{2\pi i} \int_{DE} \frac{df}{f} + \frac{1}{2\pi i} \int_{HA} \frac{df}{f} = -\frac{1}{3}v_\rho(f)$$

and

$$\frac{1}{2\pi i} \int_{FG} \frac{df}{f} = -\frac{1}{2}v_i(f).$$

Now we consider the integration

$$\frac{1}{2\pi i}\int_{EF}\frac{df}{f}+\frac{1}{2\pi i}\int_{GH}\frac{df}{f}.$$

The mapping $z\mapsto -z^{-1}$ maps EF onto GH in opposite direction and

$$f(-z^{-1})=z^k f(z).$$

So if we take the logarithmic derivative, we get

$$\frac{df(-z^{-1})}{f(-z^{-1})}=\frac{kdz}{z}+\frac{df(z)}{f(z)}.$$

Thus it follows

$$\frac{1}{2\pi i}\int_{EF}\frac{df}{f}+\frac{1}{2\pi i}\int_{GH}\frac{df}{f}=-\frac{1}{2\pi i}\int_{EF}\frac{kdz}{z}=\frac{k}{12}.$$

Consequently, we have

$$-v_\infty(f)-\frac{1}{2}v_i(f)-\frac{1}{3}v_\rho(f)+\frac{k}{12}=\sum_{p\in D}v_p(f),$$

where D is a truncated domain of $\mathcal{H}$. As D approaches $\mathcal{F}$, we get our assertion. $\qquad\square$

Corollary 4.1.2. *Let $A_k(\Gamma)$ be the vector space of modular forms of weight k. Then*

$$\dim A_k(\Gamma)=1$$

for $k=0,4,6,8,10$ with the basis 1, G_4, G_6, G_4^2 and $G_4 G_6$, respectively.

Proof. When $k=4$ and F is a nonconstant modular form of weight 4, then

$$v_\rho(F)=1\quad\text{and}\quad v_p(F)=0,\quad p\neq\rho.$$

Note that G_4 is a modular form of weight 4. If f is another modular form of weight 4, then there exists λ such that

$$v_\rho(f-\lambda G_4)\geq 2.$$

This forces $f=\lambda G_4$ since any nonconstant modular form F of weight 4 must satisfy $v_\rho(F)=1$. The remaining cases can be proved in the similar way. $\qquad\square$

Consider the *discriminant* $\Delta(z)$ given by

$$\Delta(z) = \frac{1}{1728} \left(E_4^3(z) - E_6^2(z) \right).$$

Then $v_\infty(\Delta(z)) = 1$ and $\Delta(z)$ is a cusp form of weight 12 with no other zero in the upper half-plane. The multiplication

$$f \mapsto f\Delta$$

then establishes one-to-one correspondence modular forms of weight k and cusp forms of weight $k + 12$. Also the set of cusp forms has codimension one in the set of modular forms. So that we have

$$\dim A_k(\Gamma) = \dim S_{k+12}(\Gamma) = \dim A_{k+12}(\Gamma) - 1.$$

So we have the following proposition.

Proposition 4.1.3. *The dimension of the vector space of modular forms of weight k is given as follows:*

$$\dim A_k(\Gamma) = \begin{cases} \left[\frac{k}{12} \right], & \text{if } k \equiv 2 \pmod{12}; \\ \left[\frac{k}{12} \right] + 1, & \text{if } k \equiv 0, 4, 6, 8, 10 \pmod{12}. \end{cases}$$

Corollary 4.1.4. *Notation as above. Then we have*

$$\bigcup_{k=0}^{\infty} A_k(\Gamma) = \mathbb{C}[G_4, G_6] \quad \text{and} \quad \sum_{k=0}^{\infty} \dim A_k(\Gamma) T^k = \frac{1}{(1 - T^4)(1 - T^6)}.$$

We leave this proof to reader as an exercise.

4.2. Dedekind Eta Function

The well-known *Dedekind eta function* $\eta(z)$ is defined by

$$\eta(z) = e^{\pi i z/12} \prod_{n=1}^{\infty} \left(1 - e^{2\pi i n z} \right), \quad z \in \mathscr{H}.$$

Since the infinite series

$$\sum_{n=1}^{\infty} e^{2\pi i n z}$$

converges absolutely and uniformly on any compact subset of $\mathcal{H}$, so does the infinite product appearing in $\eta(z)$ (see, for example, Ahlfors [1, p. 191]). Thus $\eta(z)$ is a holomorphic function on $\mathcal{H}$ without any zero. Here we shall prove a transformation formula of $\eta(z)$ and then conclude that $\eta^{24}(z)$ is a modular form of weight 12. Indeed, we have

$$E_4^3(z) - E_6^2(z) = 1728\eta^{24}(z).$$

Proposition 4.2.1. *The Dedekind eta function $\eta(z)$ satisfies the transformation formula*

$$\eta\left(-\frac{1}{z}\right) = \left(\frac{z}{i}\right)^{1/2}\eta(z).$$

Proof. Note that $\eta(z)$ has no zeros on $\mathcal{H}$. By taking the logarithm, the assertion is equivalent to

$$\log\eta\left(-\frac{1}{z}\right) = \log\eta(z) + \frac{1}{2}\log\left(\frac{z}{i}\right).$$

By definition, we have

$$\log\eta(z) = \frac{\pi i z}{12} + \sum_{n=1}^{\infty}\log\left(1 - e^{2\pi i n z}\right)$$
$$= \frac{\pi i z}{12} - \sum_{n=1}^{\infty}\sum_{m=1}^{\infty}\frac{e^{2\pi i m n z}}{m}.$$

Set

$$f(z) = \sum_{n=1}^{\infty}\sum_{m=1}^{\infty}\frac{e^{2\pi i m n z}}{m} = \frac{\pi i z}{12} - \log\eta(z).$$

The zeta function associated with f is given by

$$\sum_{n=1}^{\infty}\sum_{m=1}^{\infty}\frac{1}{m}(mn)^{-s} = \zeta(s)\zeta(s+1).$$

Set

$$\Lambda(f, s) = (2\pi)^{-s}\Gamma(s)\zeta(s)\zeta(s+1).$$

In light of the duplication formula

$$\Gamma\left(\frac{s}{2}\right)\Gamma\left(\frac{s+1}{2}\right) = 2^{1-s}\Gamma(s)\Gamma\left(\frac{1}{2}\right),$$

we have

$$\Lambda(f,s) = \frac{1}{2}X(s)X(1+s),$$

where

$$X(s) = \pi^{-s/2}\Gamma\left(\frac{s}{2}\right)\zeta(s).$$

The functional equation $X(1-s) = X(s)$ then implies

$$\Lambda(f,-s) = \Lambda(f,s).$$

Also $\Lambda(f,s)$ has simple poles at $s = 1$ and $s = -1$ with residue

$$\frac{\pi}{12} \quad \text{and} \quad -\frac{\pi}{12},$$

respectively. At $s = 0$, $\Lambda(f,s)$ has a pole of order 2. The function

$$\Lambda(f,s) - \frac{\pi}{12(s-1)} + \frac{\pi}{12(s+1)} + \frac{1}{2s^2}$$

is entire and bounded on any vertical strip. So the inverse Mellin transform gives

$$f(iy) = \frac{1}{2\pi i}\int_{\operatorname{Re} s=2} y^{-s}\Lambda(f,s)\,ds.$$

Now change the integral path from $\operatorname{Re} s = 2$ to $\operatorname{Re} s = -2$ by noting that $y^{-s}\Lambda(f,s)$ has poles only at $s = 1, 0, -1$ with residues $\pi/(12y)$, $\log y/2$, $-\pi y/12$, respectively. Hence

$$\begin{aligned}
f(iy) &= \frac{1}{2\pi i}\int_{\operatorname{Re} s=-2} y^{-s}\Lambda(f,s)\,ds + \frac{\pi}{12y} + \frac{1}{2}\log y - \frac{\pi y}{12} \\
&= \frac{1}{2\pi i}\int_{\operatorname{Re} s=2} y^{s}\Lambda(f,-s)\,ds + \frac{\pi}{12y} + \frac{1}{2}\log y - \frac{\pi y}{12} \\
&= f(iy^{-1}) + \frac{\pi}{12y} + \frac{1}{2}\log y - \frac{\pi y}{12}.
\end{aligned}$$

Thus the equation

$$\log\eta\left(-\frac{1}{z}\right) = \log\eta(z) + \frac{1}{2}\log\left(\frac{z}{i}\right)$$

holds for $z = iy$. Since both sides of the identity above are holomorphic functions, it holds for all $z \in \mathscr{H}$.   □

Remark 4.2.2. If we take the logarithmic derivative of $\eta(z)$, the result is

$$\frac{\eta'(z)}{\eta(z)} = \frac{2\pi i}{24}\left(1 - 24\sum_{n=1}^{\infty}\frac{ne^{2\pi inz}}{1 - e^{2\pi inz}}\right)$$

$$= \frac{\pi i}{12}E_2(z),$$

where

$$E_2(z) = 1 - 24\sum_{n=1}^{\infty}\sigma_1(n)e^{2\pi inz}$$

is defined by the series

$$\frac{1}{\zeta(2)}\sum_{(m,n)\neq(0,0)}\frac{1}{(mz+n)^2}.$$

The transformation formula for $\eta(z)$ then implies

$$z^{-2}E_2\left(-\frac{1}{z}\right) = E_2(z) + \frac{6}{\pi iz}.$$

Recall that

$$\Delta(z) = \frac{1}{1728}\left(E_4^3(z) - E_6^2(z)\right)$$

is a cusp form of weight 12. From Proposition 4.2.1, and the fact that $\eta^{24}(z+1) = \eta^{24}(z)$, we find that $\eta^{24}(z)$ is a cusp form of weight 12. As the dimension of the vector space of cusp forms of weight 12 is 1, we have the following proposition.

Proposition 4.2.3. *If*

$$\Delta(z) = \frac{1}{1728}\left(E_4^3(z) - E_6^2(z)\right),$$

then

$$\Delta(z) = e^{2\pi iz}\prod_{n=1}^{\infty}\left(1 - e^{2\pi inz}\right)^{24}.$$

The *Ramanujan's tau function* $\tau(n)$ is defined by

$$\sum_{n=1}^{\infty}\tau(n)e^{2\pi inz} = e^{2\pi iz}\prod_{n=1}^{\infty}\left(1 - e^{2\pi inz}\right)^{24}, \quad z\in\mathcal{H}.$$

The function $\tau(n)$ is multiplicative, namely,

$$\tau(mn) = \tau(m)\tau(n)$$

if m and n are relatively prime since $\Delta(z)$ is an eigenfunction of Hecke operators. Ramanujan conjectured that

$$|\tau(n)| < n^{11/2}\sigma_0(n),$$

where $\sigma_0(n)$ is the number of divisors of n. This conjecture was proved by P. Deligne as a consequence of his proof of the Weil conjecture.

4.3. Selberg Trace Formula

Another way to find the dimension of the vector space of modular forms of weight k is through the well-known *Selberg trace formula*.

Theorem 4.3.1. *Let $S_k(\Gamma)$ be the vector space of cusp forms of weight k. Then for $k > 5$, one has*

$$\dim S_k(\Gamma) = \frac{k-1}{4\pi} \int_{\Gamma\backslash\mathscr{H}} y^{k-2} \sum_{\gamma\in\bar{\Gamma}} \left[\frac{1}{2i}\left(z - \overline{\gamma(z)}\right)\right]^{-k} \overline{j(\gamma,z)}^{-k} \, dz.$$

Here we describe the general procedure to produce the trace formula and to evaluate the formula to yield a dimension.

A. How to deduce Selberg trace formula for any bounded symmetric domain?

1. General principle. Let $S_k(\Gamma)$ be the vector space of cusp forms of weight k with respect to $\Gamma = \mathrm{SL}_2(\mathbb{Z})$. It is a finite-dimensional vector space with the *Peterson inner product*

$$\langle f, g\rangle = \int_{\Gamma\backslash\mathscr{H}} y^{k-2} f(z)\overline{g(z)} \, dz.$$

So it is a Hilbert space and has an orthonormal basis $\{\varphi_n(z)\}_{n=1}^N$. Let

$$K(z_1, z_2) = \sum_{n=1}^N \varphi_n(z_1)\overline{\varphi_n(z_2)}.$$

Note also that for any f in $S_k(\Gamma)$, if we express f as a linear combination

$$\sum_{j=1}^N c_j\varphi_j(z),$$

one has

$$f(z) = \int_{\Gamma \backslash \mathscr{H}} y^{k-2} K(z, z') f(z') \, dz'.$$

Therefore, by a standard argument, we have

$$N = \dim S_k(\Gamma) = \int_{\Gamma \backslash \mathscr{H}} y^{k-2} K(z, z) \, dz.$$

This is the Selberg trace formula for the dimension of $S_k(\Gamma)$.

2. The construction of kernel functions. Let D be the unit disc on the complex plane

$$D = \left\{ w \in \mathbb{C} \mid 1 - |w|^2 > 0 \right\}.$$

Also let $L_k^1(D)$ be the set of measurable functions f on D such that

$$\int_D (1 - |w|^2)^{k-2} |f(w)| \, dw < +\infty.$$

Proposition 4.3.2. *For any holomorphic function f in $L_k^1(D)$ and any integer $k > 1$, one has*

$$f(0) = \frac{k-1}{\pi} \int_D (1 - |w|^2)^{k-2} f(w) \, dw.$$

Proof. Suppose that

$$f(w) = \sum_{n=0}^{\infty} a_n w^n.$$

Then a direct calculation yields

$$\int_D (1 - |w|^2)^{k-2} \sum_{n=0}^{\infty} a_n w^n \, dw = \int_0^1 \int_0^{2\pi} (1 - r^2)^{k-2} \sum_{n=0}^{\infty} a_n r^n e^{in\theta} \, d\theta r dr$$

$$= \frac{\pi}{k-1} a_0$$

$$= \frac{\pi}{k-1} f(0).$$

$\square$

Corollary 4.3.3. *For any holomorphic function f in $L_k^1(D)$ with $k > 1$, one has*

$$f(w_0) = \frac{k-1}{\pi} \int_D (1 - |w|^2)^{k-2} (1 - w_0 \overline{w})^{-k} f(w) \, dw.$$

Proof. Let

$$g(w) = f\left(\frac{w - w_0}{-1 + \overline{w}_0 w}\right)(1 - \overline{w}_0 w)^{-k}.$$

Then g is a holomorphic function in $L_k^1(D)$ and

$$f(w_0) = g(0)$$
$$= \frac{k-1}{\pi}\int_D (1 - |w|^2)^{k-2} g(w)\, dw$$
$$= \frac{k-1}{\pi}\int_D (1 - |w|^2)^{k-2} f\left(\frac{w - w_0}{-1 + \overline{w}_0 w}\right)(1 - \overline{w}_0 w)^{-k}\, dw$$
$$= \frac{k-1}{\pi}\int_D (1 - |w|^2)^{k-2}(1 - w_0\overline{w})^{-k} f(w)\, dw.$$

$\square$

Note that the Cayley transform

$$w = \frac{z - i}{z + i}$$

maps the upper half-plane $\mathscr{H}$ onto the unit disc D. Consequently, we have the following proposition.

Proposition 4.3.4. *For any holomorphic function f in $L_k^1(\mathscr{H})$ with $k > 1$, one has*

$$f(z_0) = \frac{k-1}{4\pi}\int_{\mathscr{H}} y^{k-2}\left[\frac{1}{2i}(z_0 - \overline{z})\right]^{-k} f(z)\, dz.$$

In particular, if f satisfies

$$f(\gamma(z)) = j(\gamma, z)^k f(z)$$

for all $\gamma \in \Gamma$, then

$$f(z_0) = \frac{k-1}{4\pi}\int_{\Gamma\backslash\mathscr{H}} y^{k-4}\sum_{\gamma\in\overline{\Gamma}}\left[\frac{1}{2i}\left(z_0 - \overline{\gamma(z)}\right)\right]^{-k}\overline{j(\gamma, z)}^{-k} f(z)\, dz.$$

For any holomorphic function f in $L_k^1(\mathscr{H}) \cap L_k^2(\mathscr{H})$, the function

$$f_\Gamma(z) = \sum_{\gamma\in\Gamma} f(\gamma(z)) j(\gamma, z)^{-k}$$

is called a Poincaré series. It is a function in $S_k(\Gamma)$ and

$$f_\Gamma(z_0) = \frac{k-1}{4\pi} \int_{\Gamma \backslash \mathscr{H}} y^{k-2} \sum_{\gamma \in \overline{\Gamma}} \left[\frac{1}{2i} \left(z_0 - \overline{\gamma(z)} \right) \right]^{-k} \overline{j(\gamma, z)}^{-k} f_\Gamma(z) \, dz.$$

Hence the function

$$\widetilde{K}(z_1, z_2) = \frac{k-1}{4\pi} \sum_{\gamma \in \overline{\Gamma}} \left[\frac{1}{2i} \left(z_1 - \overline{\gamma(z_2)} \right) \right]^{-k} \overline{j(\gamma, z_2)}^{-k}$$

is a kernel function of the vector space $S_k(\Gamma)$.

B. How to evaluate Selberg trace formula?

1. Conjugacy classes of $\Gamma = \mathrm{SL}_2(\mathbb{Z})$. An element $\gamma \in \Gamma$ is an *elliptic element* if γ has a unique fixed point on $\mathscr{H}$. So γ is conjugate in $\mathrm{SL}_2(\mathbb{R})$ to an element of the form

$$\begin{bmatrix} \cos\theta & -\sin\theta \\ \sin\theta & \cos\theta \end{bmatrix}.$$

An equivalent condition for γ to be elliptic is

$$|\operatorname{tr}(\gamma)| < 2.$$

So if $\gamma \in \Gamma$ is elliptic, then γ has the characteristic polynomial $X^2 + 1$ or $X^2 + X + 1$ or $X^2 - X + 1$.

Suppose that $\gamma \in \Gamma$, $\gamma \neq E_2$. We say that γ is a *parabolic element* if γ has a fixed point on $\mathbb{R} \cup \{\infty\}$. Here are another two equivalent conditions.

(1) The element γ is conjugate in $\mathrm{SL}_2(\mathbb{R})$ to an element of the form

$$\pm \begin{bmatrix} 1 & h \\ 0 & 1 \end{bmatrix}, \quad h \neq 0.$$

(2) $|\operatorname{tr}(\gamma)| = 2$.

An element $\gamma \in \Gamma$ is *hyperbolic* if γ has two different fixed points on $\mathbb{R} \cup \{\infty\}$. There are two other equivalent conditions.

(1) The element γ is conjugate in $\mathrm{SL}_2(\mathbb{R})$ to an element of the form

$$\pm \begin{bmatrix} r & 0 \\ 0 & 1/r \end{bmatrix}, \quad r \neq 0.$$

(2) $|\operatorname{tr}(\gamma)| > 2$.

Here we shall mention a basic property of the kernel function

$$K(z, \gamma(z)) = \frac{k-1}{4\pi} y^k \left[\frac{1}{2i} \left(z - \overline{\gamma(z)} \right) \right]^{-k} \overline{j(\gamma, z)}^{-k}.$$

Then for any $\delta \in \Gamma$,

$$\int_{\delta(\mathscr{F})} K(z, \gamma(z)) \, d\mu(z) = \int_{\mathscr{F}} K(\delta(z), \gamma\delta(z)) \, d\mu(z)$$
$$= \int_{\mathscr{F}} K(z, \delta^{-1}\gamma\delta(z)) \, d\mu(z),$$

where $d\mu(z) = dxdy/y^2$ is the invariant measure on the upper half-plane. So it follows that

$$\int_{C(\gamma)\backslash\mathscr{H}} K(z, \gamma(z)) \, d\mu(z) = \int_{\mathscr{F}} \sum_{\lambda \in \{\gamma\}} K(z, \lambda(z)) \, d\mu(z),$$

where

$$C(\gamma) = \{\delta \in \Gamma \mid \delta^{-1}\gamma\delta = \gamma\}$$

is the centralizer of γ in $\mathrm{SL}_2(\mathbb{Z})$ and

$$\{\gamma\} = \{\delta^{-1}\gamma\delta \mid \delta \in \Gamma\}$$

is the conjugacy class represented by γ.

For any elliptic element $\gamma \in \Gamma$, the set $C(\gamma)$ is a finite set. So

$$\int_{\mathscr{F}} \sum_{\lambda \in \{\gamma\}} K(z, \lambda(z)) \, d\mu(z) = \int_{C(\gamma)\backslash\mathscr{H}} K(z, \gamma(z)) \, d\mu(z)$$
$$= \frac{1}{|C(\gamma)|} \int_{\mathscr{H}} K(z, \gamma(z)) \, d\mu(z)$$
$$= \frac{1}{|C(\gamma)|} \frac{\overline{\lambda}^k}{1 - \overline{\lambda}^2},$$

where $\lambda = e^{i\theta}$ is an eigenvalue of γ. For parabolic elements of Γ, we have to collect a suitable family of conjugacy classes of parabolic elements and add the convergence factor $y^{-\epsilon}$ to the integration. For the conjugacy classes of hyperbolic elements, the contribution is zero. Here are results of the final calculation.

(a) The contribution from the identity is given by

$$\frac{k-1}{4\pi} \int_{\Gamma \backslash \mathscr{H}} \frac{dxdy}{y^2} = \frac{k-1}{4\pi} \frac{\pi}{3} = \frac{k-1}{12}.$$

(b) The contribution from the conjugacy class represented by $\begin{bmatrix} 0 & -1 \\ 1 & 0 \end{bmatrix}$ is given by

$$\frac{1}{2} \frac{i^k}{1-i^2} = \frac{i^k}{4}.$$

(c) The contribution from conjugacy class represented by $\begin{bmatrix} 0 & -1 \\ 1 & 1 \end{bmatrix}$ and $\begin{bmatrix} 0 & -1 \\ 1 & 1 \end{bmatrix}^2$ is given by

$$\frac{1}{3} \left(\frac{\rho^k}{1-\rho^2} + \frac{\bar{\rho}^k}{1-\bar{\rho}^2} \right), \quad \rho = e^{2\pi i/3}.$$

(d) The contribution from conjugacy class represented by $\pm \begin{bmatrix} 1 & n \\ 0 & 1 \end{bmatrix}$, $n \in \mathbb{Z}$, $n \neq 0$ is $-1/2$.

This is just a rough sketch of obtaining the dimension formula of cusp forms via Selberg trace formula. The details was described in the author's monograph [17].

4.4. Exercises

1. Let $A_k(\Gamma)$ be the vector space of modular forms of weight k. Prove that

$$\sum_{k=0}^{\infty} \dim A_k(\Gamma) T^k = \frac{1}{(1-T^4)(1-T^6)}.$$

2. Classify the conjugacy classes of $\mathrm{SL}_2(\mathbb{Z})$.

3. For a holomorphic function f in $L^1_k(\mathscr{H})$ and $L^2_k(\mathscr{H})$, prove that

$$f_\Gamma(z) = \sum_{\gamma \in \Gamma} f(\gamma(z)) j(\gamma, z)^{-k}$$

is a modular form of weight k and

$$\int_{\mathscr{F}} y^{k-2} |f_\Gamma(z)|^2 \, dz < +\infty.$$

4. Compute the contribution from the conjugacy class represented by

$$\begin{bmatrix} r & 0 \\ 0 & 1/r \end{bmatrix}, \quad r > 0$$

to the dimension formula for the cusp forms.

5. For any elliptic element $\gamma \in \Gamma$, let $C(\gamma)$ be the centralizer of γ in Γ and $\lambda = e^{i\theta}$ be an eigenvalue of γ. Prove that the contribution from the conjugacy class represented by γ is

$$\frac{\overline{\lambda}^k}{|C(\gamma)|} \frac{k-1}{\pi} \int_0^1 \int_0^{2\pi} \frac{(1-r^2)^{k-2}}{(1-\overline{\lambda}^2 r^2)^k} r \, dr \, d\theta.$$

Also evaluate the above integral.

Chapter 5

Bernoulli Identities and Applications

Bernoulli numbers appear as special values of zeta functions at negative integers and identities relating the Bernoulli numbers follow as a consequence of identities among the corresponding zeta functions. The most famous example is that of the special values of the Riemann zeta function and Bernoulli numbers due to Euler, namely,

$$\frac{1}{1^{2m}} + \frac{1}{2^{2m}} + \cdots + \frac{1}{n^{2m}} + \cdots = \frac{(-1)^{m-1}(2\pi)^{2m}B_{2m}}{2(2m)!}.$$

In this chapter we introduce a general principle for producing Bernoulli identities and developing their applications.

5.1. Two Classical Bernoulli Identities

Proposition 5.1.1. *For each positive integer $n \geq 2$, we have*

$$\sum_{k=1}^{n-1} \frac{(2n)!}{(2k)!(2n-2k)!} B_{2k}B_{2n-2k} = -(2n+1)B_{2n}.$$

Proof. Consider the zeta function

$$Z_1(s) = \sum_{n_1=1}^{\infty} \sum_{n_2=1}^{\infty} (n_1 + n_2)^{-s}, \quad \mathrm{Re}\, s > 2.$$

61

Let $n = n_1 + n_2$ be a new variable in place of n_2 in the summation. Then for $\operatorname{Re} s > 2$,

$$Z_1(s) = \sum_{n=2}^{\infty} n^{-s} \sum_{n_1=1}^{n-1} 1 = \sum_{n=2}^{\infty} n^{-s}(n-1) = \zeta(s-1) - \zeta(s).$$

One can now make an analytic continuation for $Z_1(s)$. In particular, one has for $n \geq 2$,

$$Z_1(2 - 2n) = \zeta(1 - 2n) - \zeta(2 - 2n) = -\frac{B_{2n}}{2n}.$$

On the other hand, for $\operatorname{Re} s > 2$, we have

$$Z_1(s)\Gamma(s) = \sum_{n_1=1}^{\infty} \sum_{n_2=1}^{\infty} \int_0^{\infty} t^{s-1} e^{-(n_1+n_2)t}\, dt$$

$$= \int_0^{\infty} t^{s-1} \sum_{n_1=1}^{\infty} \sum_{n_2=1}^{\infty} e^{-(n_1+n_2)t}\, dt$$

$$= \int_0^{\infty} t^{s-1} \left(\frac{1}{e^t - 1}\right)^2 dt.$$

Then using a general process as we had done for $\zeta(s)$, we find that

$$Z_1(2 - 2n) = (-1)^{2n-2}(2n - 2)! \times \text{the coefficient of } t^{2n}$$

$$\text{in the Taylor expansion at } t = 0 \text{ of } \left(\frac{t}{e^t - 1}\right)^2$$

$$= (2n - 2)! \sum_{k=0}^{2n} \frac{B_k B_{2n-k}}{k!(2n - k)!}$$

$$= \sum_{k=0}^{n} \frac{(2n - 2)!}{(2k)!(2n - 2k)!} B_{2k} B_{2n-2k}.$$

Comparing two expressions for $Z_1(2 - 2n)$, we obtain our assertion.   □

Remark 5.1.2. The identity in Proposition 5.1.1 is equivalent to

$$\sum_{k=1}^{n-1} \zeta(2k)\zeta(2n - 2k) = \frac{2n + 1}{2}\zeta(2n).$$

Here we provide a second proof. Consider the function

$$F(t) = \frac{t^2}{e^t - 1}.$$

First expanding $F(t)$ in power series and then differentiating the series term by term, we get

$$F'(t) = \sum_{n=0}^{\infty} \frac{(n+1)B_n t^n}{n!}.$$

On the other hand, we have

$$\begin{aligned}
F'(t) &= \frac{2t}{e^t - 1} - \frac{t^2 e^t}{(e^t - 1)^2} \\
&= \frac{2t}{e^t - 1} - \frac{t^2}{e^t - 1} - \left(\frac{t}{e^t - 1}\right)^2 \\
&= \sum_{n=0}^{\infty} \frac{2B_n t^n}{n!} - \sum_{n=0}^{\infty} \frac{B_n t^{n+1}}{n!} - \left(\sum_{n=0}^{\infty} \frac{B_n t^n}{n!}\right)^2.
\end{aligned}$$

Comparing the coefficients of t^{2n} in two different power series expansions of $F'(t)$, we get

$$\sum_{k=1}^{n-1} \frac{B_{2k} B_{2n-2k}}{(2k)!(2n-2k)!} = -(2n+1)\frac{B_{2n}}{(2n)!}.$$

This is equivalent to the identity given in Proposition 5.1.1.

Proposition 5.1.3. *For each positive integer $n \geq 4$, one has*

$$\sum_{k=2}^{n-2} \frac{(2n-2)!}{(2k-2)!(2n-2k-2)!} \frac{B_{2k}}{2k} \frac{B_{2n-2k}}{2n-2k} = \left(-\frac{B_{2n}}{2n}\right) \frac{(2n+1)(2n-6)}{6(2n-2)(2n-3)}.$$

Proof. Consider the zeta function

$$Z_2(s) = \sum_{n_1=1}^{\infty} \sum_{n_2=1}^{\infty} n_1 n_2 (n_1 + n_2)^{-s}, \quad \text{Re } s > 4.$$

Again, let $n = n_1 + n_2$ be a new variable in place of n_2. Then for $\text{Re } s > 4$,

$$\begin{aligned}
Z_2(s) &= \sum_{n=2}^{\infty} n^{-s} \sum_{n_1=1}^{n-1} (n-n_1)n_1 \\
&= \frac{1}{6} \sum_{n=2}^{\infty} n^{-s}(n^3 - n) \\
&= \frac{1}{6} \{\zeta(s-3) - \zeta(s-1)\}.
\end{aligned}$$

One can now make an analytic continuation for $Z_2(s)$. In particular, one has for $n \geq 4$,

$$Z_2(4 - 2n) = \frac{1}{6} \left\{ \zeta(1 - 2n) - \zeta(3 - 2n) \right\}$$

$$= \frac{1}{6} \left(-\frac{B_{2n}}{2n} + \frac{B_{2n-2}}{2n - 2} \right).$$

On the other hand, we have

$$Z_2(s)\Gamma(s) = \int_0^\infty t^{s-1} \left(\sum_{n=1}^\infty n e^{-nt} \right)^2 dt$$

and hence

$$Z_2(4 - 2n) = (-1)^{2n-4}(2n - 4)! \times \text{the coefficient of } t^{2n-4}$$

$$\text{in the asymptotic expansion at } t = 0 \text{ of } \left(\sum_{n=1}^\infty n e^{-nt} \right)^2.$$

Note that

$$\sum_{n=1}^\infty e^{-nt} = \frac{1}{e^t - 1} = \frac{1}{t} + \sum_{n=1}^\infty \frac{B_n t^{n-1}}{n!}, \quad |t| < 2\pi.$$

Differentiating the above identity with respect to t, we get

$$\sum_{n=1}^\infty n e^{-nt} = \frac{1}{t^2} - \sum_{n=2}^\infty \frac{B_n}{n} \frac{t^{n-2}}{(n - 2)!}.$$

Consequently, we get

$$Z_2(4 - 2n)$$

$$= (2n - 4)! \left\{ -\frac{2B_{2n}}{2n(2n - 2)!} + \sum_{k=1}^{n-1} \frac{B_{2k}}{2k(2k - 2)!} \frac{B_{2n-2k}}{(2n - 2k)(2n - 2k - 2)!} \right\}.$$

Our assertion then follows from two different expressions of $Z_2(4-2n)$. $\square$

Remark 5.1.4. The identity of Proposition 5.1.3 appeared in [45] is a consequence of an identity among Eisenstein series of different weights:

$$\frac{1}{6}(2n+1)(2n-1)(2n-6)G_{2n}(z) = \sum_{k=2}^{n-2}(2k-1)(2n-2k-1)G_{2k}(z)G_{2n-2k}(z),$$

where $G_{2k}(z)$ is the Eisenstein series defined by

$$G_{2k}(z) = \sum_{(m,n)\neq(0,0)} (m+nz)^{-2k}.$$

To produce more Bernoulli identities of similar kind, we have to consider more zeta functions of the form

$$\sum_{n_1=1}^{\infty}\sum_{n_2=1}^{\infty}\cdots\sum_{n_r=1}^{\infty} (n_1+n_2+\cdots+n_r)^{-s}, \quad \mathrm{Re}\, s > r$$

or

$$\sum_{n_1=1}^{\infty}\sum_{n_2=1}^{\infty}\cdots\sum_{n_r=1}^{\infty} n_1^{\alpha_1} n_2^{\alpha_2}\cdots n_r^{\alpha_r}(n_1+n_2+\cdots+n_r)^{-s}, \quad \mathrm{Re}\, s > r + |\alpha|,$$

where $|\alpha| = \alpha_1 + \alpha_2 + \cdots + \alpha_r$. However, we need a convenient way to evaluate these zeta functions at negative integers. Here we mention some useful ones.

Proposition 5.1.5. *Suppose that $a_1, a_2, \ldots, a_r$ are positive real numbers and let*

$$Z_3(s) = \sum_{n_1=1}^{\infty}\sum_{n_2=1}^{\infty}\cdots\sum_{n_r=1}^{\infty} (a_1 n_1 + a_2 n_2 + \cdots + a_r n_r)^{-s}, \quad \mathrm{Re}\, s > r.$$

Then the zeta function defined as above has its analytic continuation in the whole complex plane and for each positive integer $m \geq r$,

$$Z_3(r-m) = (-1)^{m-r}(m-r)! \sum_{|\alpha|=m} a_1^{\alpha_1-1} a_2^{\alpha_2-1}\cdots a_r^{\alpha_r-1}\frac{B_{\alpha_1} B_{\alpha_2}\cdots B_{\alpha_r}}{\alpha_1!\alpha_2!\cdots\alpha_r!}.$$

Proof. It follows from for $\mathrm{Re}\, s > r$,

$$Z_3(s)\Gamma(s) = \int_0^{\infty} \frac{t^{s-1}}{(e^{a_1 t}-1)(e^{a_2 t}-1)\cdots(e^{a_r t}-1)}\, dt$$

and hence

$$Z_3(r-m) = (-1)^{m-r}(m-r)! \times \text{ the coefficient of } t^m$$

in the Taylor expansion at $t=0$ of the function

$$\left(\frac{t}{e^{a_1 t}-1}\right)\left(\frac{t}{e^{a_2 t}-1}\right)\cdots\left(\frac{t}{e^{a_r t}-1}\right).$$

$\square$

Proposition 5.1.6. *For positive numbers* $a_1, a_2, \ldots, a_r$ *and* $x_1, x_2, \ldots, x_r$, *let*

$$Z_4(s) = \sum_{n_1=0}^{\infty} \sum_{n_2=0}^{\infty} \cdots \sum_{n_r=0}^{\infty} [a_1(n_1 + x_1) + \cdots + a_r(n_r + x_r)]^{-s}, \quad \text{Re } s > r.$$

Then $Z_4(s)$ *has its analytic continuation in the whole complex plane. Furthermore, for each positive integer* $m \geq r$,

$$Z_4(r - m)$$
$$= (-1)^r (m - r)! \sum_{|\alpha|=m} a_1^{\alpha_1 - 1} a_2^{\alpha_2 - 1} \cdots a_r^{\alpha_r - 1} \frac{B_{\alpha_1}(x_1) B_{\alpha_2}(x_2) \cdots B_{\alpha_r}(x_r)}{\alpha_1! \alpha_2! \cdots \alpha_r!}.$$

Proof. It follows from

$$Z_4(r - m) = (-1)^{m-r} (m - r)! \times \text{the coefficient of } t^m$$

in the Taylor expansion at $t = 0$ of the function

$$\prod_{j=1}^{r} \frac{t e^{a_j(1 - x_j)t}}{e^{a_j t} - 1}.$$

$\square$

5.2. Zeta Functions Associated with Linear Forms

The evaluation of zeta functions associated with linear forms plays the most important role in our production of Bernoulli identities. So it is worthwhile paying attention to this topic. Let $L(X) = a_1 x_1 + a_2 x_2 + \cdots + a_r x_r + \delta$ be a linear form with positive coefficients a_j, $j = 1, 2, \ldots, r$, $\delta \geq 0$ and $\beta = (\beta_1, \beta_2, \ldots, \beta_r)$ be r-tuple of nonnegative integers. For $\text{Re } s > |\beta| + r$, consider the zeta function

$$Z(L, \beta; s) = \sum_{n_1=1}^{\infty} \sum_{n_2=1}^{\infty} \cdots \sum_{n_r=1}^{\infty} n_1^{\beta_1} n_2^{\beta_2} \cdots n_r^{\beta_r} (a_1 n_1 + a_2 n_2 + \cdots + a_r n_r + \delta)^{-s}.$$

The special value of $Z(L, \beta; s)$ at the negative integer $s = -m$ is given by

$$Z(L, \beta; -m) = (-1)^m m! \times \text{the coefficient of } t^m$$

$$\text{in the asymptotic expansion at } t = 0 \text{ of}$$

$$G(t) = \prod_{j=1}^{r} \left(\sum_{n=1}^{\infty} n^{\beta_j} e^{-a_j nt} \right) e^{-\delta t}.$$

Differentiating both sides of

$$\sum_{n=1}^{\infty} e^{-nt} = \frac{1}{t} + \sum_{n=1}^{\infty} \frac{B_n t^{n-1}}{n!}$$

with respect to t β_j-times, we obtain

$$\sum_{n=1}^{\infty} n^{\beta_j} e^{-nt} = \frac{\beta_j!}{t^{\beta_j+1}} + (-1)^{\beta_j} \sum_{n \geq \beta_j+1} \frac{B_n}{n} \frac{t^{n-\beta_j-1}}{(n-\beta_j-1)!},$$

and hence

$$\sum_{n=1}^{\infty} n^{\beta_j} e^{-a_j nt} = \frac{\beta_j!}{(a_j t)^{\beta_j+1}} + (-1)^{\beta_j} \sum_{n \geq \beta_j+1} \frac{B_n}{n} \frac{(a_j t)^{n-\beta_j-1}}{(n-\beta_j-1)!}.$$

To express $Z(L, \beta; -m)$ more precisely, we introduce new mappings from the polynomial ring $\mathbb{C}[X_1, X_2, \ldots, X_\ell]$ to $\mathbb{C}$. Let J^ℓ, $0 \leq \ell \leq r$, be the linear extension of the mapping

$$X_1^{\alpha_1} \cdots X_\ell^{\alpha_\ell} \mapsto \zeta(-\alpha_1) \cdots \zeta(-\alpha_\ell) = \frac{(-1)^\alpha B_{\alpha_1+1} \cdots B_{\alpha_\ell+1}}{(\alpha_1+1) \cdots (\alpha_\ell+1)}$$

from $\mathbb{C}^\ell[X_1, X_2, \ldots, X_\ell]$ into $\mathbb{C}$ for $\ell \geq 1$ and $J^0(c) = 0$. For example, we have

$$J(x^m) = (-1)^m \frac{B_{m+1}}{m+1}.$$

Proposition 5.2.1. [16] *Let $Z(L, \beta; s)$ be the zeta function defined by*

$$Z(L, \beta; s) = \sum_{n_1=1}^{\infty} \sum_{n_2=1}^{\infty} \cdots \sum_{n_r=1}^{\infty} n_1^{\beta_1} n_2^{\beta_2} \cdots n_r^{\beta_r} (a_1 n_1 + a_2 n_2 + \cdots + a_r n_r + \delta)^{-s},$$

where $L = a_1 n_1 + a_2 n_2 + \cdots + a_r n_r + \delta$. Then $Z(L, \beta; s)$ has its analytic continuation in the complex plane. In particular, for each positive integer m,

$$Z(L, \beta; -m) = J^r \left(X^\beta L^m(X) \right)$$

$$+ \sum_{j_1 \leq \cdots \leq j_\ell} J^{r-\ell} \left(\int_{\Delta(j_1, \ldots, j_\ell)} X^\beta L^m(X) \, dX_{j_1} \cdots dX_{j_\ell} \right),$$

where $\{j_1, \ldots, j_\ell\}$ ranges over all nonempty subset of the set $I = \{1, 2, \ldots, r\}$ in the summation and $\Delta(j_1, \ldots, j_\ell)$ is the simplex in $\mathbb{R}^\ell$ defined by

$$X_{j_1} \leq 0, \quad X_{j_2} \leq 0, \quad \ldots, \quad X_{j_\ell} \leq 0, \quad L(X) \geq 0.$$

Proof. See [16] for the details of the proof. $\qquad\qquad\qquad\qquad\qquad\square$

Here we give an example to illustrate the application of the above theorem.

Proposition 5.2.2. *For each positive integer $n \geq 4$, we have*

$$\sum_{k=3}^{n-3} \frac{(2n-6)!}{(2k-3)!(2n-2k-3)!} \frac{B_{2k}}{2k} \frac{B_{2n-2k}}{2n-2k}$$

$$= \frac{(2n+1)(4n^2 - 26n + 60)}{30(2n-3)(2n-4)(2n-5)} \left(-\frac{B_{2n}}{2n} \right) + \frac{1}{60} B_{2n-4}.$$

Proof. Consider the zeta function

$$Z_5(s) = \sum_{n_1=1}^{\infty} \sum_{n_2=1}^{\infty} n_1^2 n_2^2 (n_1 + n_2)^{-s}, \quad \operatorname{Re} s > 6.$$

By a change of variable in the summation, we get

$$Z_5(s) = \frac{1}{30} \left\{ \zeta(s-5) - \zeta(s-1) \right\}.$$

So the special value at $s = 6 - 2n$ is equal to

$$-\frac{1}{30} \frac{B_{2n}}{2n} + \frac{1}{30} \frac{B_{2n-4}}{2n-4}.$$

On the other hand, we have

$$Z_5(6 - 2n) = J^2 \left(x^2 y^2 (x + y)^{2n-6}\right) + J \left(\int_0^{-y} x^2 y^2 (x + y)^{2n-6} \, dx\right)$$

$$+ J \left(\int_0^{-x} x^2 y^2 (x + y)^{2n-6} \, dy\right)$$

$$= \sum_{k=2}^{n-2} \frac{(2n - 6)!}{(2k - 3)!(2n - 2k - 3)!} \frac{B_{2k}}{2k} \frac{B_{2n-2k}}{2n - 2k}$$

$$+ \frac{4}{(2n - 3)(2n - 4)(2n - 5)} \frac{B_{2n}}{2n}.$$

Thus our assertion then follows by comparing two different expressions of $Z_5(6 - 2n)$. $\square$

As another example, we consider the zeta function

$$Z_6(s) = \sum_{n_1=1}^{\infty} \sum_{n_2=1}^{\infty} \sum_{n_3=1}^{\infty} n_1 n_2 n_3 (n_1 + n_2 + n_3)^{-s}$$

which is equal to

$$\frac{1}{120} \zeta(s - 5) - \frac{1}{24} \zeta(s - 3) + \frac{1}{30} \zeta(s - 1)$$

since

$$\sum_{n_1+n_2+n_3=n} n_1 n_2 n_3 = \frac{1}{120} \left(n^5 - 5n^3 + 4n\right).$$

By setting $s = 6 - 2n$ with $n \geq 6$, we get

$$Z_6(6 - 2n) = J^3 \left(xyz(x + y + z)^{2n-6}\right)$$

$$+ 3J^2 \left(\int_0^{-(x+y)} xyz(x + y + z)^{2n-6} \, dx\right)$$

$$+ 3J \left(\iint_D xyz(x + y + z)^{2n-6} \, dxdy\right),$$

where D is a domain in $\mathbb{R}^n$ with $x + y + z \geq 0$, $x \leq 0$, $y \leq 0$. In the final, we get the identity

$$\sum_{\substack{p+q+r=n \\ p,q,r \geq 2}} \frac{(2n - 6)!}{(2p - 2)!(2q - 2)!(2r - 2)!} \frac{B_{2p} B_{2q} B_{2r}}{8pqr}$$

$$= \left(-\frac{B_{2n}}{2n}\right) \left(\frac{1}{120} - \frac{2n^2 - 5n}{(2n - 2)(2n - 3)(2n - 4)(2n - 5)}\right) + \frac{1}{80} \left(\frac{B_{2n-4}}{2n - 4}\right).$$

5.3. Zeta Functions Associated with Rational Functions

Let $P(T)$ be a polynomial function in T and $n_1, n_2, \ldots, n_r$ be positive integers. Consider the rational function

$$F(T) = \frac{P(T)}{(1 - T^{n_1})(1 - T^{n_2}) \cdots (1 - T^{n_r})}.$$

Suppose that at $T = 0$, $F(T)$ has a Taylor expansion

$$\sum_{n=0}^{\infty} a(n) T^n.$$

The zeta function associated with $F(T)$ is given by

$$Z_F(s) = \sum_{n=1}^{\infty} a(n) n^{-s}, \operatorname{Re} s > r.$$

Indeed for $\operatorname{Re} s > r$, we have

$$Z_F(s) \Gamma(s) = \int_0^{\infty} t^{s-1} \left\{ F(e^{-t}) - F(0) \right\} dt.$$

So $Z_F(s)$ has its analytic continuation with a general procedure as we have done before for $\zeta(s)$. Furthermore, for any positive integer m,

$$Z_F(-m) = (-1)^m m! \times \text{the coefficient of } t^m$$
$$\text{in the asymptotic expansion at } t = 0 \text{ of } F(e^{-t}).$$

This provides a nice way to evaluate $Z_F(s)$ at negative integers.

However, there are always alternating way to express $Z_F(s)$ as a finite sum of shifted Riemann zeta functions or Hurwitz zeta functions as we shall see. This leads to Bernoulli identities with a polynomial function in Bernoulli numbers on one side and a finite sum of Bernoulli numbers or Bernoulli polynomials on the other side.

Here we mention a few well-known examples.

Example 5.3.1. Consider the rational function

$$F(T) = \frac{1}{(1 - T)^2}.$$

Then $F(T)$ has a Taylor expansion

$$F(T) = \sum_{k=0}^{\infty} (k+1)T^k,$$

so that for $\operatorname{Re} s > 2$,

$$Z_F(s) = \sum_{k=1}^{\infty} (k+1)k^{-s} = \zeta(s-1) + \zeta(s).$$

On the other hand, we have

$$F(T) = \sum_{n_1=0}^{\infty} \sum_{n_2=0}^{\infty} T^{n_1+n_2}$$

and hence

$$Z_F(s) = \sum_{n_1=1}^{\infty} \sum_{n_2=1}^{\infty} (n_1 + n_2)^{-s} + 2\zeta(s).$$

Consequently, we have the identity

$$\sum_{n_1=1}^{\infty} \sum_{n_2=1}^{\infty} (n_1 + n_2)^{-s} = \zeta(s-1) - \zeta(s).$$

Example 5.3.2. Consider the rational function

$$F(T) = \frac{T^2}{(1-T)^4}.$$

The Taylor expansion at $T = 0$ of $F(T)$ is given by

$$\frac{1}{6} \sum_{k=0}^{\infty} (k+1)(k+2)(k+3)T^{k+2},$$

so that

$$Z_F(s) = \frac{1}{6} \left\{ \zeta(s-3) - \zeta(s-1) \right\}.$$

On the other hand, we have

$$Z_F(s) = \sum_{m_1=0}^{\infty} \sum_{m_2=0}^{\infty} \sum_{m_3=0}^{\infty} \sum_{m_4=0}^{\infty} (m_1 + m_2 + m_3 + m_4 + 2)^{-s}$$

$$= \sum_{n_1=1}^{\infty} \sum_{n_2=1}^{\infty} n_1 n_2 (n_1 + n_2)^{-s},$$

where $n_1 = m_1 + m_2 + 1$ and $n_2 = m_3 + m_4 + 1$.

Example 5.3.3. Consider the zeta function

$$F(T) = \frac{1}{(1 - T^2)(1 - T^3)}.$$

The zeta function associated with $F(T)$ is given by

$$Z_F(s) = \sum_{n_1=1}^{\infty} \sum_{n_2=1}^{\infty} (2n_1 + 3n_2)^{-s} + (2^{-s} + 3^{-s})\zeta(s).$$

Decomposing $F(T)$ into a sum of its partial fractions as

$$F(T) = \frac{1}{6(1 - T)^2} + \frac{1}{4(1 + T)} + \frac{3T^2 - T + 7}{12(1 - T^3)},$$

we obtain another expression of $Z_F(s)$ as

$$Z_F(s) = \frac{1}{6}\zeta(s - 1) + \frac{1}{6}\zeta(s) + \frac{2^{-s}}{4}\zeta(s) - \frac{2^{-s}}{4}\zeta\left(s; \frac{1}{2}\right)$$

$$+ \frac{3^{-s}}{4}\zeta\left(s; \frac{2}{3}\right) - \frac{3^{-s}}{12}\zeta\left(s; \frac{1}{3}\right) + \frac{7}{12}3^{-s}\zeta(s).$$

In general, we are able to decompose $F(T)$ of the particular forms into a sum of its partial fractions so that the zeta functions associated with each partial fraction is a sum of shifted Riemann zeta functions or Hurwitz zeta functions. Indeed we can still achieve such results even without writing $F(T)$ as partial fractions. We demonstrate such a procedure by an example.

Example 5.3.4. Let p and q be different positive integers. Consider the rational function

$$F(T) = \frac{1}{(1 - T^p)(1 - T^q)}.$$

The zeta function associated with $F(T)$ is given by

$$Z_F(s) = \sum_{n_1=1}^{\infty} \sum_{n_2=1}^{\infty} (pn_1 + qn_2)^{-s} + (p^{-s} + q^{-s})\zeta(s).$$

Setting $s = 2 - 2n$ with $n \geq 2$, we get

$$Z_F(2 - 2n) = \sum_{k=0}^{n} \frac{(2n - 2)!}{(2k)!(2n - 2k)!} p^{2k-1} q^{2n-2k-1} B_{2k} B_{2n-2k}.$$

On the other hand, we have

$$F(T) = \frac{(1 + T^p + \cdots + T^{p(q-1)})(1 + T^q + \cdots + T^{q(p-1)})}{(1 - T^{pq})^2}$$

$$= \sum_{k=0}^{\infty} (k+1)(1 + T^p + \cdots + T^{p(q-1)})(1 + T^q + \cdots + T^{q(p-1)})T^{pqk}$$

so that

$$Z_F(s) = (pq)^{-s} \left[\zeta(s-1) + \zeta(s) \right] + (pq)^{-s} \sum_{j,k} \zeta\left(s-1; \frac{j}{p} + \frac{k}{q} \right)$$

$$+ (pq)^{-s} \sum_{j,k} \left(1 - \frac{j}{p} - \frac{k}{q} \right) \zeta\left(s; \frac{j}{p} + \frac{k}{q} \right),$$

where $0 \le j \le p-1$, $0 \le k \le q-1$, but $(j,k) \ne (0,0)$. Consequently, we have

$$Z_F(2 - 2n) = (pq)^{2n-2} \left(-\frac{B_{2n}}{2n} \right) - \frac{(pq)^{2n-2}}{2n} \sum_{j,k} B_{2n} \left(\frac{j}{p} + \frac{k}{q} \right)$$

$$- \frac{(pq)^{2n-2}}{2n-1} \sum_{j,k} \left(1 - \frac{j}{p} - \frac{k}{q} \right) B_{2n-1} \left(\frac{j}{p} + \frac{k}{q} \right).$$

Theorem 5.3.5. *For a pair of positive integers p, q and any positive integer $n \ge 2$, we have*

$$\sum_{k=1}^{n-1} \frac{(2n)!}{(2k)!(2n-2k)!} p^{2k} q^{2n-2k} B_{2k} B_{2n-2k}$$

$$= -(p^n + q^n + (pq)^{2n-1})(2n-1)B_{2n}$$

$$- (pq)^{2n-1}(2n-1) \sum_{j,k} B_{2n} \left(\frac{j}{p} + \frac{k}{q} \right)$$

$$- (pq)^{2n-1} 2n \sum_{j,k} \left(1 - \frac{j}{p} - \frac{k}{q} \right) B_{2n-1} \left(\frac{j}{p} + \frac{k}{q} \right).$$

As our final example, we consider the zeta function associated with the rational function

$$F(T) = \frac{1}{(1 - T^4)(1 - T^6)}.$$

Proposition 5.3.6. *For any positive integer $n \geq 2$, one has*

$$\sum_{k=1}^{n-1} \frac{(2n)!}{(2k)!(2n-2k)!} 4^{2k} 6^{2n-2k} B_{2k} B_{2n-2k}$$

$$= -\left[(2n-1)2^{2n} + 6^{2n} + 4^{2n}\right] B_{2n} + (16n)6^{2n-2} B_{2n-1}\left(\frac{1}{3}\right).$$

Proof. Consider the zeta function

$$Z_7(s) = \sum_{n_1=1}^{\infty} \sum_{n_2=1}^{\infty} (4n_1 + 6n_2)^{-s} + (4^{-s} + 6^{-s})\zeta(s)$$

which is equal to the Dirichlet series

$$\sum_{k=1}^{\infty} a(k)k^{-s},$$

where $a(k)$ is the number of nonnegative integral solutions of the equation $4m + 6n = k$. Obviously, $a(k) = 0$ if k is odd. For a positive even integer k, we have

$$a(k) = \frac{k+5}{12} + \frac{1}{4}\delta_1(k) + \frac{1}{3}\delta_2(k)$$

with

$$\delta_1(k) = \begin{cases} 1, & \text{if } k \equiv 0 \pmod{4}; \\ -1, & \text{if } k \equiv 2 \pmod{4} \end{cases}$$

and

$$\delta_2(k) = \begin{cases} 1, & \text{if } k \equiv 0 \pmod{6}; \\ -1, & \text{if } k \equiv 2 \pmod{6}; \\ 0, & \text{if } k \equiv 4 \pmod{6}. \end{cases}$$

Indeed, $a(k)$ is the dimension of the vector space of modular forms of weight k. So we have

$$Z_7(s) = \frac{1}{12}\zeta(s-1) + \frac{5}{12}\zeta(s) + 4^{-s}\zeta(s) - 4^{-s}\zeta\left(s; \frac{1}{2}\right)$$

$$+ 6^{-s}\zeta(s) - 6^{-s}\zeta\left(s; \frac{1}{3}\right).$$

Setting $s = 2 - 2n$, we get our assertion. $\qquad\square$

In the following, we shall consider a kind of function of the form

$$\frac{P(T)}{(1 - T^{m_1})(1 - T^{m_2}) \cdots (1 - T^{m_r})},$$

where $P(T)$ is not necessary a polynomial in T but a finite sum of real powers in T. The numbers m_j, $j = 1, 2, \ldots, r$, are not necessary positive integers but positive real numbers.

Proposition 5.3.7. *For each positive integer m and a real number x such that $0 \leq x < 1$, we have*

$$\sum_{n=1}^{\infty} \frac{\cos(2n\pi x)}{n^{2m}} = \frac{(-1)^{m-1}(2\pi)^{2m} B_{2m}(x)}{2(2m)!}.$$

Proof. Consider the function

$$F(T) = \frac{T^x}{1 - T}.$$

The zeta function associated with $F(T)$ is a Hurwitz zeta function and we have

$$\zeta(1 - 2m; x) = -\frac{1}{2m} B_{2m}(x) = -(2m - 1)! \frac{1}{2\pi i} \int_{|z|=\epsilon} \frac{z^{-2m} e^{-xz}}{1 - e^{-z}} \, dz.$$

Let

$$F(z) = \frac{z^{-2m} e^{-xz}}{1 - e^{-z}}$$

and C_N be the contour of rectangle with vertices $(2N + 1) + i(2N + 1)$, $(2N + 1) - i(2N + 1)$, $-(2N + 1) + i(2N + 1)$, $-(2N + 1) - i(2N + 1)$. It is easy to see that

$$\lim_{N \to \infty} \int_{C_N} F(z) \, dz = 0.$$

So it follows that

$$\zeta(1 - 2m; x) = (2m - 1)! \sum_{k \neq 0} \mathrm{Res}\,(F(z), z = 2k\pi i)$$

$$= (2m - 1)! \sum_{k=1}^{\infty} \left(\frac{1}{2k\pi i}\right)^{2m} \left(e^{-2k x \pi i} + e^{2k x \pi i}\right)$$

$$= 2(2m - 1)! \sum_{k=1}^{\infty} \frac{(-1)^m \cos(2k\pi x)}{(2k\pi)^{2m}}.$$

Thus our assertion follows. $\qquad\square$

Let α, $\beta > 0$ with $\alpha\beta = \pi^2$ and n be a positive integer. The following identity is well-known:

$$2^{2n} \sum_{k=0}^{n+1} \frac{B_{2n+2-2k}}{(2n+2-2k)!} \frac{B_{2k}}{(2k)!} \alpha^{n+1-k}(-\beta)^k$$

$$= -\alpha^{-n} \left\{ \frac{1}{2}\zeta(2n+1) + \sum_{k=1}^{\infty} \frac{k^{-2n-1}}{e^{2\alpha k}-1} \right\}$$

$$+ (-\beta)^{-n} \left\{ \frac{1}{2}\zeta(2n+1) + \sum_{k=1}^{\infty} \frac{k^{-2n-1}}{e^{2\beta k}-1} \right\}.$$

Here we shall give a new proof of this and derive some of its generalizations.

Proof. For any given $\epsilon > 0$, we consider the zeta function

$$Z_\epsilon(s) = \sum_{n_1=1}^{\infty} \sum_{n_2=1}^{\infty} \left[\sqrt{\alpha} n_1 + \left(\epsilon + i\sqrt{\beta}\right) n_2 \right]^{-s}, \quad \operatorname{Re} s > 2.$$

Note that $Z_\epsilon(s)$ has its analytic continuation and its special value at $s = -2n$ is given by

$$Z_\epsilon(-2n) = \frac{(2n)!}{\sqrt{\alpha}\left(\epsilon + i\sqrt{\beta}\right)} \sum_{k=0}^{n+1} \frac{B_{2n+2-2k}}{(2n+2-2k)!} \frac{B_{2k}}{(2k)!} \alpha^{n+1-k} \left(\epsilon + i\sqrt{\beta}\right)^{2k}.$$

On the other hand, we set

$$F_\epsilon(t) = \sum_{n_1=1}^{\infty} \sum_{n_2=1}^{\infty} e^{-\sqrt{\alpha}n_1 - (\epsilon+i\sqrt{\beta})n_2} = \frac{1}{\left(e^{\sqrt{\alpha}t}-1\right)\left(e^{(\epsilon+i\sqrt{\beta})t}-1\right)}.$$

Then we also have

$$Z_\epsilon(-2n) = (2n)! \frac{1}{2\pi i} \int_{|z|=\delta} z^{-(2n+1)} F_\epsilon(z)\, dz.$$

As $\epsilon \to 0$, we obtain

$$\frac{1}{\sqrt{\alpha\beta}i} \sum_{k=0}^{n+1} \frac{B_{2n+2-2k}}{(2n+2-2k)!} \frac{B_{2k}}{(2k)!} \alpha^{n+1-k}(-\beta)^k = \frac{1}{2\pi i} \int_{|z|=\delta} z^{-2n-1} F_0(z)\, dz.$$

The above contour integral is also equal to

$$-\sum_{k\neq 0} \operatorname{Res}\left(z^{-2n-1}F_0(z), z = \frac{2k\pi i}{\sqrt{\alpha}}, \frac{2k\pi}{\sqrt{\beta}} \right),$$

or

$$-\sum_{k\neq 0}\left(\frac{1}{2k\sqrt{\alpha}}\right)^{2n+1}\frac{1}{\sqrt{\beta}i(e^{2k\alpha}-1)}+\sum_{k\neq 0}\left(\frac{1}{2k\sqrt{\beta}i}\right)^{2n+1}\frac{1}{\sqrt{\alpha}(e^{2k\beta}-1)}.$$

An elementary calculation then yields our conclusion. $\qquad\square$

If we consider the zeta function

$$\sum_{n_1=0}^{\infty}\sum_{n_2=0}^{\infty}\left[\sqrt{\alpha}(n_1+u)+\left(\epsilon+i\sqrt{\beta}\right)(n_2+v)\right]^{-s},\quad \operatorname{Re}s>2$$

instead, we obtain the following Bernoulli identities.

Proposition 5.3.8. *For all positive numbers α, β with $\alpha\beta=\pi^2$ and real numbers u, v such that $0\leq u,v\leq 1$, we have*

$$2^{2n}\sum_{k=0}^{n+1}\frac{B_{2n+2-2k}(u)}{(2n+2-2k)!}\frac{B_{2k}(v)}{(2k)!}\alpha^{n-k+1}(-\beta)^k$$

$$=-\frac{1}{2}\alpha^{-n}\sum_{k=1}^{\infty}\frac{k^{-2n-1}\cos(2k\pi v)\left[e^{2ku\alpha}+e^{2k(1-u)\alpha}\right]}{e^{2k\alpha}-1}$$

$$+\frac{1}{2}(-\beta)^{-n}\sum_{k=1}^{\infty}\frac{k^{-2n-1}\cos(2k\pi u)\left[e^{2k(1-v)\beta}+e^{2kv\beta}\right]}{e^{2k\beta}-1}$$

and

$$2^{2n}\sum_{k=0}^{n}\frac{B_{2n-2k+1}(u)}{(2n-2k+1)!}\frac{B_{2k+1}(v)}{(2k+1)!}\alpha^{n-k+1/2}\beta^{k+1/2}(-1)^k$$

$$=-\frac{1}{2}\alpha^{-n}\sum_{k=1}^{\infty}\frac{k^{-2n-1}\sin(2k\pi v)\left[e^{2ku\alpha}-e^{2k(1-u)\alpha}\right]}{e^{2k\alpha}-1}$$

$$+\frac{1}{2}(-\beta)^{-n}\sum_{k=1}^{\infty}\frac{k^{-2n-1}\sin(2k\pi u)\left[e^{2k(1-v)\beta}-e^{2kv\beta}\right]}{e^{2k\beta}-1}.$$

5.4. Kummer's Congruences

Recall that Bernoulli numbers B_n, $n=0,1,2,\ldots$, defined by

$$\frac{t}{e^t-1}=\sum_{n=0}^{\infty}\frac{B_n t^n}{n!},\quad |t|<2\pi$$

are rational numbers. As the function

$$F(t) = \frac{t}{e^t - 1} + \frac{t}{2}$$

is an even function, i.e., $F(-t) = F(t)$, this implies that all Bernoulli numbers of odd index, except for B_1, are equal to zero. In other words,

$$B_{2m+1} = 0 \quad \text{for } m \geq 1.$$

There is a theorem concerning the denominator of Bernoulli numbers of even index.

Theorem 5.4.1 (von Staudt-Clausen Theorem). [8] *Suppose that p is a prime and m is a positive even integer. If $p - 1$ is not a divisor of m, then p is not a divisor of denominator of B_m. If $p - 1$ is a divisor of m, then pB_m is p-integral and*

$$pB_m \equiv -1 \pmod{p}.$$

The classical Kummer's congruences asserted as follows [8, p. 384]:

Theorem 5.4.2 (Kummer's Congruences). *Suppose that p is a prime and m is a positive even integer. If $p - 1$ is not a divisor of m, then the number B_m/m is p-integral and*

$$\frac{B_{m+p-1}}{m+p-1} \equiv \frac{B_m}{m} \pmod{p}.$$

Remark 5.4.3. Kummer's congruences play important roles in the p-adic interpolation of the classical Riemann zeta function. Consider the zeta function

$$\zeta_p(s) = (1 - p^{-s})\zeta(s) = \sum_{\substack{n=1 \\ (n,p)=1}}^{\infty} n^{-s}, \quad \operatorname{Re} s > 1.$$

The special value at $s = 1 - m$ is given by

$$\zeta_p(1 - m) = -(1 - p^{m-1})\frac{B_m}{m}.$$

An extension of Kummer's congruence asserted that if

$$m \equiv n \pmod{(p-1)p^N}$$

then

$$\left(1 - p^{m-1}\right) \frac{B_m}{m} \equiv \left(1 - p^{n-1}\right) \frac{B_n}{n} \pmod{p^{N+1}}.$$

So it tells us that $\zeta_p(s)$ is a continuous function on the ring of p-adic integers $\mathbb{Z}_p$, that is,

$$\zeta_p(1 - m) \equiv \zeta_p(1 - n) \pmod{p^{N+1}}$$

if $m \equiv n \pmod{(p-1)p^N}$.

Here we are going to prove the above extension of Kummer's congruences.

Proposition 5.4.4. *For any prime number p and complex number s with* $\operatorname{Re} s > 1$, *one has*

$$\left(1 - p^{-s}\right) \zeta(s) = p^{-(N+1)s} \sum_{\substack{1 \leq j < p^{N+1} \\ (j,p)=1}} \zeta\left(s; \frac{j}{p^{N+1}}\right).$$

Proof. Consider the zeta function $Z_F(s)$ associated with the rational function

$$F(T) = \frac{1}{1-T} - \frac{1}{1-T^p}.$$

It is easy to see that for $\operatorname{Re} s > 1$,

$$Z_F(s) = \sum_{k=1}^{\infty} k^{-s} - \sum_{k=1}^{\infty} (kp)^{-s} = \left(1 - p^{-s}\right) \zeta(s).$$

On the other hand, we have for any nonnegative integer N,

$$\begin{aligned}
F(T) &= \frac{T + T^2 + \cdots + T^{p-1}}{1 - T^p} \\
&= \frac{(T + T^2 + \cdots + T^{p-1})(1 + T^p + T^{2p} + \cdots + T^{p(p^N - 1)})}{1 - T^{p^{N+1}}} \\
&= \sum_{\substack{1 \leq j < p^{N+1} \\ (j,p)=1}} \sum_{k=0}^{\infty} T^{j + kp^{N+1}}.
\end{aligned}$$

It follows that for $\operatorname{Re} s > 1$,

$$Z_F(s) = p^{-(N+1)s} \sum_{\substack{1 \leq j < p^{N+1} \\ (j,p)=1}} \zeta\left(s; \frac{j}{p^{N+1}}\right).$$

Note that $Z_F(s)$ is determined by $F(T)$ uniquely through the integral formula

$$Z_F(s)\Gamma(s) = \int_0^\infty t^{s-1} F(e^{-t})\, dt, \quad \operatorname{Re} s > 1.$$

Thus our identity follows.   □

As an immediate consequence, we have the following.

Proposition 5.4.5. *Suppose that m is a positive even integer and p is an odd prime such that m is not divisible by $p - 1$. Then*

$$\left(1 - p^{m-1}\right)\frac{B_m}{m} \equiv C_0(m) + C_1(m) \pmod{p^{N+1}},$$

where

$$C_\ell(m) = \frac{1}{m} \sum_{\substack{1 \le j < p^{N+1} \\ (j,p)=1}} j^{m-\ell}\binom{m}{\ell} B_\ell p^{(N+1)(\ell-1)}, \quad 0 \le \ell \le m.$$

Proof. Setting $s = 1 - m$ in the identity in Proposition 5.4.4, we find that

$$\left(1 - p^{m-1}\right)\frac{B_m}{m} = \sum_{\ell=0}^{m} C_\ell(m).$$

Now we shall prove that $C_\ell(m) \equiv 0 \pmod{p^{N+1}}$ for $\ell \ge 2$. Note that the exponent of p that occurs in $\ell!$ is not greater than

$$\frac{\ell}{p} + \frac{\ell}{p^2} + \cdots + \frac{\ell}{p^k} + \cdots = \frac{\ell}{p-1} \le \frac{\ell}{4}$$

if $p \ge 5$. Using Theorem 5.4.1, we know that

$$C_\ell(m) = (m-1)(m-2)\cdots(m-\ell+1) \sum_{\substack{1 \le j < p^{N+1} \\ (j,p)=1}} j^{m-\ell}\frac{1}{\ell!}(pB_\ell)p^{(N+1)(\ell-1)-1}.$$

Thus $C_\ell(m) \equiv 0 \pmod{p^{N+1}}$ is true provided that

$$-\frac{\ell}{4} + (N+1)(\ell-1) - 1 \ge N + 1.$$

This is equivalent to

$$(N+1)(\ell-2) \ge \frac{\ell}{4} + 1.$$

But N is a nonnegative integer, the equality holds provided that

$$\ell - 2 \geq \frac{\ell}{4} + 1.$$

This is equivalent to $\ell \geq 4$. It follows that

$$\left(1 - p^{m-1}\right) \frac{B_m}{m} \equiv C_0(m) + C_1(m) + C_2(m) \pmod{p^{N+1}}.$$

So it suffices to prove $C_2(m) \equiv 0 \pmod{p^{N+1}}$. Note that

$$C_2(m) = \frac{m-1}{2} B_2 \sum_{(j,p)=1} j^{m-1} p^{N+1} = \frac{m-1}{12} p^{N+1} \sum_{(j,p)=1} j^{m-1}.$$

As $p = 3$ is excluded in our assumption, so

$$\frac{m}{12} p^{N+1}$$

is p-integral and it is divisible by p^{N+1}. $\square$

As shown in the above, Kummer's congruences are equivalent to

$$C_0(m) + C_1(m) \equiv C_0(n) + C_1(n) \pmod{p^{N+1}}$$

if $m \equiv n \pmod{(p-1)p^N}$. However,

$$C_1(m) = -\frac{1}{2} \sum_{\substack{1 \leq j < p^{N+1} \\ (j,p)=1}} j^{m-1}.$$

So it is easy to see that

$$C_1(m) \equiv C_1(n) \pmod{p^{N+1}}$$

since

$$j^{m-1} \equiv j^{n-1} \pmod{p^{N+1}}$$

for all integers j relatively prime to p. Consequently, Kummer's congruences are equivalent to

$$\frac{1}{mp^{N+1}} \sum_{\substack{1 \leq j < p^{N+1} \\ (j,p)=1}} j^m \equiv \frac{1}{np^{N+1}} \sum_{\substack{1 \leq j < p^{N+1} \\ (j,p)=1}} j^n \pmod{p^{N+1}}$$

if $m \equiv n \pmod{(p-1)p^N}$. To simplify the notation we write $\sum_{(j,p)=1}$ instead of

$$\sum_{\substack{1 \leq j < p^{N+1} \\ (j,p)=1}}.$$

Proposition 5.4.6. *Suppose that m, n are positive even integers and p is an odd prime with m not divisible by $p-1$. Then*

$$\frac{1}{mp^{N+1}} \sum_{(j,p)=1} j^m \equiv \frac{1}{np^{N+1}} \sum_{(j,p)=1} j^n \pmod{p^{N+1}}$$

if $m \equiv n \pmod{(p-1)p^N}$.

Proof. For an odd prime p, the multiplicative group $G = \left(\mathbb{Z}/p^{N+1}\mathbb{Z}\right)^*$ is cyclic [36], so there exists a generator g of G and hence

$$C_0(m) = \frac{1}{mp^{N+1}} \left(1^m + g^m + g^{2m} + \cdots + g^{m((p-1)p^N - 1)} \right)$$

$$= \frac{1}{mp^{N+1}} \frac{g^{m(p-1)p^N} - 1}{g^m - 1}.$$

The order of g is $(p-1)p^N$. Suppose that

$$g^{(p-1)p^N} = 1 + kp^{N+1},$$

then it is a direct verification that

$$\frac{g^{m(p-1)p^N} - 1}{mp^{N+1}}$$

is p-integral and

$$\frac{1}{mp^{N+1}} \left(g^{m(p-1)p^N} - 1 \right) \equiv k \pmod{p^{N+1}}$$

by an induction on $\nu_p(m)$, the largest nonnegative integer k such that p^k is a divisor of m. Therefore, our assertion is equivalent to

$$\frac{k}{g^m - 1} \equiv \frac{k}{g^n - 1} \pmod{p^{N+1}}$$

if $m \equiv n \pmod{(p-1)p^N}$.

As $p - 1$ is not a divisor of m, both $g^m - 1$ and $g^n - 1$ are invertible elements of G and

$$g^m - 1 \equiv g^n - 1 \pmod{p^{N+1}}.$$

So that

$$(g^m - 1)^{-1} \equiv (g^n - 1)^{-1} \pmod{p^{N+1}}. \qquad \square$$

A similar argument leads to Kummer's congruences on Bernoulli polynomials.

Theorem 5.4.7. [27] *Suppose that $p \geq 5$ is an odd prime and m, n are positive integers such that $m - 1$ is not a divisor of n. Then for any positive integer k relatively prime to p and nonnegative integers α, β such that $\alpha + jk = p\beta$ for some j with $0 \leq j \leq p - 1$, one has*

$$\frac{1}{m}\left[B_m\left(\frac{\alpha}{k}\right) - p^{m-1}B_m\left(\frac{\beta}{k}\right)\right]$$
$$\equiv \frac{1}{n}\left[B_n\left(\frac{\alpha}{k}\right) - p^{n-1}B_n\left(\frac{\beta}{k}\right)\right] \pmod{p^{N+1}}$$

if $m \equiv n \pmod{(p-1)p^N}$.

Proof. To prove the theorem, we consider the rational function

$$F(T) = \frac{T^\alpha}{1 - T^k} - \frac{T^{p\beta}}{1 - T^{kp}}.$$

Rewriting $F(T)$ as a rational function with the denominator $1 - T^{kp^{N+1}}$, we get the identity for $\mathrm{Re}\, s > 1$,

$$\zeta\left(s; \frac{\alpha}{k}\right) - p^{-s}\zeta\left(s; \frac{\beta}{k}\right) = p^{-(N+1)s} \sum_{j \equiv \alpha \pmod{k}} \zeta\left(s; \frac{j}{kp^{N+1}}\right),$$

where j ranges over all integers of the form $\alpha + k\ell$ and $\alpha \leq j < \alpha + kp^{N+1}$, $(j, p) = 1$. Setting $s = 1 - m$ and then modulo p^{N+1}, we get

$$\frac{1}{m}\left[B_m\left(\frac{\alpha}{k}\right) - p^{m-1}B_m\left(\frac{\beta}{k}\right)\right]$$
$$\equiv \frac{1}{mk^m p^{N+1}} \sum_{j \equiv \alpha \pmod{k}} j^m - \frac{1}{2k^{m-1}} \sum_{j \equiv \alpha \pmod{k}} j^{m-1} \pmod{p^{N+1}}.$$

As the mapping $x \mapsto kx + \alpha$ is one-to-one from $G = (\mathbb{Z}/p^{N+1}\mathbb{Z})^*$ into itself. So the rest of the proof is the same as we have done before. $\qquad \square$

5.5. Exercises

1. Compute the special values at negative integers of the following zeta functions:

   (a) $\displaystyle\sum_{n_1=1}^{\infty}\sum_{n_2=1}^{\infty}\sum_{n_3=1}^{\infty}(n_1+n_2+n_3)^{-s}$;

   (b) $\displaystyle\sum_{n_1=0}^{\infty}\sum_{n_2=0}^{\infty}\sum_{n_3=0}^{\infty}\sum_{n_4=0}^{\infty}(n_1+n_2+n_3+n_4+2)^{-s}$.

2. Prove that for each positive integer $n \geq 3$,

$$\sum_{\substack{p+q+r=n \\ p,q,r\geq 1}}\frac{(2n)!}{(2p)!(2q)!(2r)!}B_{2p}B_{2q}B_{2r}$$

$$=\frac{(2n+1)(2n+2)}{2}B_{2n}+\frac{n(2n-1)}{2}B_{2n-2}.$$

3. Prove that for each positive integer $n \geq 4$,

$$\sum_{\substack{p+q+r+\ell=n \\ p,q,r,\ell\geq 1}}\frac{(2n)!}{(2p)!(2q)!(2r)!(2\ell)!}B_{2p}B_{2q}B_{2r}B_{2\ell}$$

$$=-\left\{\frac{(2n+1)(2n+2)(2n+3)}{6}B_{2n}+\frac{4n^2(2n+1)}{3}B_{2n-2}\right\}.$$

4. Prove that

$$\sum_{p+q=2n+1}\frac{(2n+1)!}{p!q!}B_p(x)B_q(x)$$

$$=-2nB_{2n+1}(2x)+2(2n+1)B_1(x)B_{2n}(2x).$$

5. Prove that for a positive integer $n \geq 2$,

$$\sum_{k=1}^{2n-1}\frac{(2n)!}{k!(2n-k)!}B_k(x)B_{2n-k}(y)$$

$$=-(2n-1)B_{2n}(x+y)-2n(1-x-y)B_{2n-1}(x+y)-B_{2n}(x)$$

$$-B_{2n}(y).$$

Chapter 6

Euler Sums and Recent Development

For a pair of positive integers p and q with $q > 1$, the classical Euler sum is defined as

$$S_{p,q} = \sum_{k=1}^{\infty} \frac{1}{k^q} \sum_{j=1}^{k} \frac{1}{j^p}.$$

The number $w = p+q$ is called the weight of $S_{p,q}$. The evaluations of Euler sums in terms of values at positive integers of Riemann zeta function has a long history. It was first proposed in 1742 in a letter from Goldbach to Euler. In 1775, Euler proved the case $p = 1$ and gave a general formula for $S_{p,q}$ when the weight is odd without any proof. N. Nielsen [43] was the first to fill in the gap by giving the correct version and a proof.

6.1. Two Different Approaches for Evaluations

By the definition of Euler sums, we see immediately that

$$S_{p,q} + S_{q,p} = \zeta(p)\zeta(q) + \zeta(p+q)$$

for p, $q \geq 2$. This is called the reflection formula of Euler sums.

Theorem 6.1.1 (Euler). *For each positive integer n with $n \geq 2$, we have*

$$S_{1,n} = \frac{n+2}{2}\zeta(n+1) - \frac{1}{2}\sum_{r=2}^{n-1}\zeta(r)\zeta(n+1-r).$$

Proof. Rewrite $S_{1,n}$ as

$$S_{1,n} = \zeta(n+1) + \sum_{k=1}^{\infty}\sum_{j=1}^{\infty}\frac{1}{j(k+j)^n}.$$

In light of the partial fraction decomposition

$$\frac{1}{X(X+T)^n} = \frac{1}{T^n}\left(\frac{1}{X} - \frac{1}{X+T}\right) - \sum_{r=2}^{n}\frac{1}{T^{n+1-r}(X+T)^r},$$

we get

$$\sum_{k=1}^{\infty}\sum_{j=1}^{\infty}\frac{1}{j(k+j)^n} = \sum_{k=1}^{\infty}\frac{1}{k^n}\sum_{j=1}^{\infty}\left(\frac{1}{j} - \frac{1}{k+j}\right)$$

$$-\sum_{r=2}^{n}\sum_{k=1}^{\infty}\sum_{j=1}^{\infty}\frac{1}{k^{n+1-r}(k+j)^r}$$

$$= \sum_{k=1}^{\infty}\frac{1}{k^n}\sum_{j=1}^{k}\frac{1}{j} - \sum_{r=2}^{n}\{S_{n+1-r,r} - \zeta(n+1)\}$$

$$= (n-1)\zeta(n+1) - \sum_{r=2}^{n-1}S_{n+1-r,r}.$$

By the reflection formula, we have

$$\sum_{r=2}^{n-1}S_{n+1-r,r} = \frac{1}{2}\sum_{r=2}^{n-1}\{S_{n+1-r,r} + S_{r,n+1-r}\}$$

$$= \frac{1}{2}\sum_{r=2}^{n-1}\{\zeta(r)\zeta(n+1-r) + \zeta(n+1)\}.$$

Thus our assertion follows by an easy calculation.   □

Remark 6.1.2. If we employ the more general formula for $w = p + q$,

$$\frac{1}{X^p(X+T)^q} = \sum_{\ell=2}^{p}(-1)^{p+\ell}\binom{w-\ell-1}{q-1}\frac{1}{T^{w-\ell}X^\ell}$$
$$+ (-1)^p \sum_{\ell=2}^{q}\binom{w-\ell-1}{p-1}\frac{1}{T^{w-\ell}(X+T)^\ell}$$
$$+ (-1)^{p+1}\binom{w-2}{p-1}\frac{1}{T^{w-1}}\left(\frac{1}{X}-\frac{1}{X+T}\right),$$

we get the identity

$$S_{p,q} - \zeta(w) = \sum_{\ell=2}^{p}(-1)^{p+\ell}\binom{w-\ell-1}{q-1}\zeta(w-\ell)\zeta(\ell)$$
$$+ (-1)^p \sum_{\ell=2}^{q}\binom{w-\ell-1}{p-1}\{S_{w-\ell,\ell} - \zeta(w)\}$$
$$+ (-1)^{p+1}\binom{w-2}{p-1}S_{1,w-1}.$$

It is interesting to note that Euler sums of odd weight come from integral transforms of products of Bernoulli polynomials.

Proposition 6.1.3. *For all positive integers m and n, we have*

$$\int_0^1 B_{2m}(x)B_{2n+1}(x)\cot(\pi x)\,dx$$
$$= \frac{(-1)^{m+n}4(2m)!(2n+1)!}{(2\pi)^{2m+2n+1}}\left\{S_{2m,2n+1} - \frac{1}{2}\zeta(2m+2n+1)\right\}.$$

Proof. Replace $B_{2n+1}(x)$ by its Fourier expansion

$$\frac{(-1)^{n+1}2(2n+1)!}{(2\pi)^{2n+1}}\sum_{k=1}^{\infty}\frac{\sin(2k\pi x)}{k^{2n+1}}$$

and employ the identity

$$\sin(2k\pi x)\cot(\pi x) = 1 + 2\sum_{j=1}^{k}\cos(2j\pi x) - \cos(2k\pi x),$$

it suffices to evaluate

$$\int_0^1 B_{2m}(x)\,dx \quad \text{and} \quad \int_0^1 B_{2m}(x)\cos(2j\pi x)\,dx.$$

The antiderivative of $B_{2m}(x)$ is $B_{2m+1}(x)/(2m+1)$, so

$$\int_0^1 B_{2m}(x)\,dx = \frac{1}{2m+1}\left\{B_{2m+1}(1) - B_{2m+1}(0)\right\} = 0.$$

On the other hand, we replace $B_{2m}(x)$ by its Fourier expansion

$$\frac{(-1)^{m-1}2(2m)!}{(2\pi)^{2m}} \sum_{k=1}^{\infty} \frac{\cos(2k\pi x)}{k^{2m}},$$

then

$$\int_0^1 B_{2m}(x)\cos(2j\pi x)\,dx$$
$$= \frac{(-1)^{m-1}2(2m)!}{(2\pi)^{2m}} \sum_{\ell=1}^{\infty} \frac{1}{\ell^{2m}} \int_0^1 \cos(2\ell\pi x)\cos(2j\pi x)\,dx$$
$$= \frac{(-1)^{m-1}(2m)!}{(2\pi)^{2m}} \frac{1}{j^{2m}}.$$

Thus our assertion follows.   □

The following proposition follows with a similar argument.

Proposition 6.1.4. *For each positive integer n, we have*

$$\int_0^1 B_{2n+1}(x)\cot(\pi x)\,dx = \frac{(-1)^{n+1}2(2n+1)!}{(2\pi)^{2n+1}}\zeta(2n+1).$$

Next we are going to produce a kind of Bernoulli identities with products of Bernoulli polynomials on one side and linear combinations of Bernoulli polynomials on the other side. To do so, we consider a product of two Hurwitz zeta functions given by

$$\zeta(ps;x)\zeta(qs;x) = \sum_{n_1=0}^{\infty} \sum_{n_2=0}^{\infty} [(n_1+x)^p(n_2+x)^q]^{-s}, \quad \text{Re}\,s > 1, x > 0.$$

Separate the above double series into 2 subseries according to $n_1 \geq n_2$ or $n_1 \leq n_2$. When $n_1 \geq n_2$, we let $n_1 = n_1' + n_2$. When $n_1 \leq n_2$, we let

$n_2 = n_1 + n_2'$. This leads to the identity

$$\zeta(ps; x)\zeta(qs; x) = \sum_{n_1=0}^{\infty} \sum_{n_2=0}^{\infty} [(n_1 + n_2 + x)^p (n_2 + x)^q]^{-s}$$

$$+ \sum_{n_1=0}^{\infty} \sum_{n_2=0}^{\infty} [(n_1 + x)^p (n_1 + n_2 + x)^q]^{-s}$$

$$- \zeta(ps + qs; x).$$

A new kind of zeta functions appears on the right-hand side of the above identity. The following proposition is the main tool to evaluate all these zeta functions at negative integers.

Proposition 6.1.5. *For all positive integers p, q and real numbers a, x, y with $a > 0$, $x \geq 0$ and $y > 0$, we let*

$$Z_{p,q}(s; a, x, y) = \sum_{n_1=0}^{\infty} \sum_{n_2=0}^{\infty} [(a(n_1 + x) + n_2 + y)^p (n_2 + y)^q]^{-s}, \quad \mathrm{Re}\, s > 1.$$

Then $Z_{p,q}(s; a, x, y)$ has its analytic continuation in the whole complex plane as a function of s. Furthermore, for each positive integer m and $w = pm + qm + 2$, we have

$$Z_{p,q}(-m; a, x, y) = \frac{1}{pm+1} \sum_{r=0}^{pm+1} \binom{pm+1}{r} a^{r-1} B_r(x) \frac{B_{w-r}(y)}{w-r}$$

$$+ \frac{p(-1)^{qm}}{p+q}(pm)!(qm)! a^{w-1} \frac{B_w(x)}{w!}.$$

Proof. By the definition of gamma function

$$\Gamma(s) = \int_0^{\infty} t^{s-1} e^{-t}\, dt, \quad \mathrm{Re}\, s > 0 \tag{6.1.1}$$

and with a change of variable, we get for $\mathrm{Re}\, s > 0$,

$$(a(n_1 + x) + n_2 + y)^{-ps}\, \Gamma(ps) = \int_0^{\infty} t_1^{ps-1} e^{-[a(n_1+x)+n_2+y]t_1}\, dt_1 \tag{6.1.2}$$

and

$$(n_2 + y)^{-qs} \Gamma(qs) = \int_0^{\infty} t_2^{qs-1} e^{-(n_2+y)t_2}\, dt_2. \tag{6.1.3}$$

Multiplying (6.1.2) and (6.1.3) together and letting n_1 and n_2 range over all nonnegative integers, we get for $\operatorname{Re} s > 1$,

$$Z_{p,q}(s; a, x, y)\Gamma(ps)\Gamma(qs)$$

$$= \int_0^\infty \int_0^\infty t_1^{ps-1} t_2^{qs-1} \sum_{n_1=0}^\infty \sum_{n_2=0}^\infty e^{-[a(n_1+x)+n_2+y]t_1 - (n_2+y)t_2} \, dt_1 dt_2$$

$$= \int_0^\infty \int_0^\infty t_1^{ps-1} t_2^{qs-1} \frac{e^{(1-x)at_1}}{e^{at_1} - 1} \frac{e^{(1-y)(t_1+t_2)}}{e^{t_1+t_2} - 1} \, dt_1 dt_2$$

after an exchange of the order of summation and integration. By changing of variables with

$$t_1 = tu, \quad t_2 = tv, \quad t > 0, \quad u > 0, \quad v > 0, \quad u + v = 1.$$

We get for $\operatorname{Re} s > 1$,

$$Z_{p,q}(s; a, x, y)\Gamma(s) = \int_0^\infty t^{ps+qs-3} F(t, s) \, dt \tag{6.1.4}$$

with

$$F(t, s) = \frac{\Gamma(s)}{\Gamma(ps)\Gamma(qs)} \int_0^1 u^{ps-1} v^{qs-1} \frac{te^{(1-x)atu}}{e^{atu} - 1} \frac{te^{(1-y)t}}{e^t - 1} \, du.$$

With the same argument as before, the special value at negative integer $s = -m$ of $Z_{p,q}$ depends only on the coefficient of $t^{pm+qm+2}$ of Taylor expansion at $t = 0$ of $F(t, -m)$. Indeed, for $w = pm + qm + 2$, we have

$$Z_{p,q}(-m; a, x, y) = \frac{(-1)^m m!}{p+q} \sum_{r=0}^w \frac{a^{r-1}}{r!(w-r)!} B_r(1-x) B_{w-r}(1-y) G_r(-m), \tag{6.1.5}$$

where for $\operatorname{Re} s > 0$,

$$G_r(s) = \frac{\Gamma(s)}{\Gamma(ps)\Gamma(qs)} \int_0^1 u^{ps-1} v^{qs-1} u^{r-1} \, du$$

$$= \frac{\Gamma(s)\Gamma(ps+r-1)}{\Gamma(ps)\Gamma(ps+qs+r-1)}.$$

Note that the gamma function $\Gamma(s)$ has simple poles at nonpositive integers. Then we have that $G_r(-m) = 0$ unless $0 \le r \le pm+1$ or $r = pm+qm+2$. For these exceptional cases, we first have for $0 \le r \le pm+1$,

$$G_r(-m) = (-1)^{pm+qm+m}(p+q)\frac{(pm)!(pm+qm+1-r)!}{m!(pm+1-r)!}.$$

On the other hand, when $r = pm + qm + 2$, we have

$$G_r(-m) = (-1)^{pm+m}p\frac{(pm)!(qm)!}{m!}.$$

Thus our assertion then follows from (6.1.5) with the above evaluations of $G_r(-m)$ and equation

$$B_n(1-x) = (-1)^n B_n(x). \qquad \square$$

With the help of above proposition, we get the identity for $w = 2m + 2n + 1$,

$$B_{2m}(x)B_{2n+1}(x) = 2m \sum_{k=0}^{n} \binom{2n+1}{2k} B_{2k}\frac{B_{w-2k}(x)}{w-2k}$$
$$+ (2n+1) \sum_{k=0}^{m} \binom{2m}{2k} B_{2k}\frac{B_{w-2k}(x)}{w-2k}.$$

Multiplying both sides by $\cot(\pi x)$ and then integrating x from 0 to 1, we get

$$S_{2m,2n+1} = \frac{1}{2}\zeta(w) + \sum_{k=0}^{m} \binom{w-2k-1}{2n}\zeta(2k)\zeta(w-2k)$$
$$+ \sum_{k=0}^{n} \binom{w-2k-1}{2m-1}\zeta(2k)\zeta(w-2k).$$

Here $\zeta(0) = -1/2$.

Theorem 6.1.6. *For an odd weight $w = p + q$ with $p, q \geq 2$, we have*

$$S_{p,q} = \frac{1}{2}\zeta(w) + \frac{1-(-1)^p}{2}\zeta(p)\zeta(q)$$
$$+ (-1)^p \sum_{k=0}^{[p/2]} \binom{w-2k-1}{q-1}\zeta(2k)\zeta(w-2k)$$
$$+ (-1)^p \sum_{k=0}^{[q/2]} \binom{w-2k-1}{p-1}\zeta(2k)\zeta(w-2k).$$

Other alternating Euler sums are defined as follows:

$$S_{p,q}^{+-} = \sum_{k=1}^{\infty} \frac{(-1)^{k+1}}{k^q} \sum_{j=1}^{k} \frac{1}{j^p},$$

$$S_{p,q}^{-+} = \sum_{k=1}^{\infty} \frac{1}{k^q} \sum_{j=1}^{k} \frac{(-1)^{j+1}}{j^p}$$

and

$$S_{p,q}^{--} = \sum_{k=1}^{\infty} \frac{(-1)^{k+1}}{k^q} \sum_{j=1}^{k} \frac{(-1)^{j+1}}{j^p}.$$

All these Euler sums also come from integral transforms of products of Bernoulli polynomials.

Proposition 6.1.7. *For all positive integers m and n, we have*

$$\int_0^1 B_{2m}(x) B_{2n+1}(x) \tan(\pi x)\, dx$$
$$= \frac{(-1)^{m+n} 4(2m)!(2n+1)!}{(2\pi)^{2m+2n+1}} \left\{ -S_{2m,2n+1}^{--} + \frac{1}{2}\zeta(2m+2n+1) \right\}.$$

Proof. It follows from the same procedure as in Proposition 6.1.3 except we employ the elementary identity

$$\sin(2k\pi x)\tan(\pi x) = (-1)^{k-1} + 2\sum_{j=1}^{k}(-1)^{k+j-1}\cos(2j\pi x) + \cos(2k\pi x)$$

in the first step. $\qquad\qquad\qquad\qquad\qquad\qquad\qquad\qquad\qquad\qquad \square$

Proposition 6.1.8. *For all positive integers m and n, we have*

$$\int_0^{1/2} B_{2m}\left(x + \frac{1}{2}\right) B_{2n+1}(x)\cot(\pi x)\, dx$$
$$= \frac{(-1)^{m+n+1} 2(2m)!(2n+1)!}{(2\pi)^{2m+2n+1}} \left\{ S_{2m,2n+1}^{-+} - \frac{1}{2}\eta(2m+2n+1) \right\},$$

where

$$\eta(s) = \sum_{k=1}^{\infty} \frac{(-1)^{k+1}}{k^s}, \qquad \mathrm{Re}\, s > 0.$$

Proof. With the same first step as in Proposition 6.1.3, it suffices to evaluate

$$\int_0^{1/2} B_{2m}\left(x + \frac{1}{2}\right) dx \quad \text{and} \quad \int_0^{1/2} B_{2m}\left(x + \frac{1}{2}\right)\cos(2j\pi x)\, dx.$$

The first integral is zero and

$$\int_0^{1/2} B_{2m}\left(x+\frac{1}{2}\right)\cos(2j\pi x)\,dx$$

$$= \frac{(-1)^{m-1}2(2m)!}{(2\pi)^{2m}}\sum_{\ell=1}^{\infty}\frac{1}{\ell^{2m}}\int_0^{1/2}\cos(2\ell\pi x+\ell\pi x)\cos(2j\pi x)\,dx$$

$$= \frac{(-1)^{m-1}(2m)!}{(2\pi)^{2m}}\frac{1}{2}\frac{(-1)^j}{j^{2m}}.$$

$\square$

6.2. Analogue of Euler Sums

There are many interesting generalizations for the classical Euler sums. Here we mention a few among them.

6.2.1. Extended Euler sums [13]

For positive integers p, q and k with $q > 1$, we define

$$E_{p,q}^{(k)} = \sum_{n=1}^{\infty}\frac{1}{n^q}\sum_{j=1}^{kn}\frac{1}{j^p}\quad\text{and}\quad T_{p,q}^{(k)} = \sum_{n=1}^{\infty}\frac{1}{n^q}\sum_{j=1}^{[n/k]}\frac{1}{j^p},$$

where $[x]$ is the greatest integer less than or equal to x. Note that for p, $q > 1$, we have the reflection formula

$$E_{p,q}^{(k)} + T_{q,p}^{(k)} = \zeta(p)\zeta(q) + k^{-p}\zeta(p+q).$$

To obtain the evaluations of $E_{p,q}^{(k)}$ and $T_{p,q}^{(k)}$ when $p+q$ is odd, we have to produce Bernoulli identities of $B_{2n+1}(\{kx\})B_{2m}(x)$ and $B_{2n+1}(x)B_{2m}(\{kx\})$ on one side and linear combinations of Bernoulli polynomials on the other side. So we consider a product of two Hurwitz zeta functions given by

$$\zeta(ps;kx)\zeta(qs;x) = \sum_{n_1=0}^{\infty}\sum_{n_2=0}^{\infty}[(n_1+kx)^p(n_2+x)^q]^{-s},\quad \mathrm{Re}\,s > 1, x > 0.$$

Here p, q and k are positive integers. We decompose such a zeta function into $k+1$ zeta functions considered in Proposition 6.1.5 according to $n_1 \geq kn_2$ or $n_1 < kn_2$ with details as given below.

1. When $n_1 \geq kn_2$, we simply let $n_1 = n_1' + kn_2$ with n_1' and n_2 ranging over all nonnegative integers.

2. When $n_1 < kn_2$, we let $n_1 = kn_1' + (k - j)$ with $1 \leq j \leq k$ and $n_2 = n_1' + n_2' + 1$ with new dummy variables n_1' and n_2' ranging over all nonnegative integers.

This leads to the identity

$$\zeta(ps; kx)\zeta(qs; x)$$
$$= k^{-ps} \sum_{n_1=0}^{\infty} \sum_{n_2=0}^{\infty} \left[(k^{-1}n_1 + n_2 + x)^p \, (n_2 + x)^q \right]^{-s}$$
$$+ k^{-ps} \sum_{j=1}^{k} \sum_{n_1=0}^{\infty} \sum_{n_2=0}^{\infty} \left[\left(n_1 + 1 - \frac{j}{k} + x \right)^p (n_1 + n_2 + x + 1)^q \right]^{-s}$$

and hence the Bernoulli identity

$$B_{2n+1}(kx)B_{2m}(x)$$
$$= 2m \sum_{r=0}^{n} \binom{2n+1}{2r} k^{2n-2r+1} B_{2r} \frac{B_{w-2r}(x)}{w - 2r}$$
$$+ (2n+1)k^{2n} \sum_{r=0}^{m} \binom{2m}{2r} B_{2r} \frac{B_{w-2r}(x)}{w - 2r}$$
$$+ (2n+1)k^{2n} \sum_{j=1}^{k-1} \sum_{r=0}^{2m} \binom{2m}{r} B_r \left(\frac{j}{k} \right) \frac{B_{w-r}(1 + x - j/k)}{w - r}.$$

Let $\chi_{[a,b]}(x)$ be the characteristic function on the interval $[a, b]$ defined by

$$\chi_{[a,b]}(x) = \begin{cases} 1, & \text{if } a \leq x \leq b; \\ 0, & \text{otherwise.} \end{cases}$$

Now using

$$B_n(x+1) - B_n(x) = nx^{n-1}$$

repeatedly, we get

$$B_n(kx) = B_n(\{kx\}) + nk^{n-1} \sum_{j=1}^{k-1} \left(x - \frac{j}{k} \right)^{n-1} \chi_{[j/k,1]}(x).$$

Now replace $B_{2n+1}(kx)$ and $B_{w-r}(1+x-j/k)$ by

$$B_{2n+1}(\{kx\}) + (2n+1)k^{2n} \sum_{j=1}^{k-1} \left(x - \frac{j}{k}\right)^{2n} \chi_{[j/k,1]}(x)$$

and

$$B_{w-r}\left(\left\{1+x-\frac{j}{k}\right\}\right) + (w-r)\left(x - \frac{j}{k}\right)^{w-r-1} \chi_{[j/k,1]}(x),$$

respectively. After some cancellation, we get a Bernoulli identity

$$B_{2n+1}(\{kx\})B_{2m}(x)$$
$$= 2m \sum_{r=0}^{n} \binom{2n+1}{2r} k^{2n-2r+1} B_{2r} \frac{B_{w-2r}(x)}{w-2r}$$
$$+ (2n+1)k^{2n} \sum_{r=0}^{m} \binom{2m}{2r} B_{2r} \frac{B_{w-2r}(x)}{w-2r}$$
$$+ (2n+1)k^{2n} \sum_{j=1}^{k-1}\sum_{r=0}^{2m} \binom{2m}{r} B_r\left(\frac{j}{k}\right) \frac{B_{w-r}(\{x-j/k\})}{w-r}.$$

We multiply each factor by $\cot(\pi x)$ and then integrate x from 0 to 1 to yield the following theorem.

Theorem 6.2.1. *For an odd weight $w = 2m + 2n + 1$ with m, n and k being positive integers, then*

$$E_{2m,2n+1}^{(k)} = \frac{1}{2}k^{-2m}\zeta(w) + \sum_{r=0}^{n} \binom{w-2r-1}{2m-1} k^{2n-2r-1}\zeta(2r)\zeta(w-2r)$$
$$+ k^{2n} \sum_{r=0}^{m} \binom{w-2r-1}{2n} \zeta(2r)\zeta(w-2r)$$
$$+ k^{2n} \sum_{j=1}^{k-1}\sum_{r=0}^{m} \binom{w-2r-1}{2n} C_{2r}\left(\frac{j}{k}\right) C_{w-2r}\left(\frac{j}{k}\right)$$
$$+ k^{2n} \sum_{j=1}^{k-1}\sum_{r=0}^{m-1} \binom{w-2r-1}{2n} S_{2r+1}\left(\frac{j}{k}\right) S_{w-2r-1}\left(\frac{j}{k}\right).$$

Here

$$C_t(x) = \sum_{n=1}^{\infty} \frac{\cos(2n\pi x)}{n^t} \quad and \quad S_t(x) = \sum_{n=1}^{\infty} \frac{\sin(2n\pi x)}{n^t}$$

are cosine and sine parts of the periodic zeta function defined by

$$E_t(x) = \sum_{n=1}^{\infty} \frac{e^{2\pi i n x}}{n^t}, \quad \mathrm{Re}\, t > 1, x \in \mathbb{R}.$$

6.2.2. Generalized Euler sums of even weight [26]

For a pair of positive integers p and q, we define

$$G_{p,q}^{+-} = \sum_{k=0}^{\infty} \frac{(-1)^k}{(2k+1)^q} \sum_{j=1}^{k} \frac{1}{j^p}$$

and

$$G_{p,q}^{--} = \sum_{k=0}^{\infty} \frac{(-1)^k}{(2k+1)^q} \sum_{j=1}^{k} \frac{(-1)^{j+1}}{j^p}.$$

When $p+q$ is even, both $G_{p,q}^{+-}$ and $G_{p,q}^{--}$ can be evaluated in terms of $\zeta(s)$,

$$\eta(s) = \sum_{k=1}^{\infty} \frac{(-1)^{k-1}}{k^s}, \quad \lambda(s) = \sum_{k=0}^{\infty} \frac{1}{(2k+1)^s} \quad \text{and} \quad L(s) = \sum_{k=0}^{\infty} \frac{(-1)^k}{(2k+1)^s}.$$

The special values at negative integers of the *Lerch zeta function* defined by

$$\eta(s; x) = \sum_{n=0}^{\infty} (-1)^n (n+x)^{-s}, \quad \mathrm{Re}\, s > 0, x > 0$$

are given by

$$\eta(-m; x) = \frac{E_m(x)}{2},$$

where $E_n(x)$, $n = 0, 1, 2, 3, \ldots$, are *Euler polynomials* defined by

$$\frac{2e^{xt}}{e^t + 1} = \sum_{n=0}^{\infty} \frac{E_n(x)t^n}{n!}, \quad |t| < \pi.$$

For $0 \le x < 1$, we have the Fourier expansion for $E_n(x)$ as

$$E_{2n-1}(x) = \frac{(-1)^n 4(2n-1)!}{\pi^{2n}} \sum_{k=0}^{\infty} \frac{\cos(2k+1)\pi x}{(2k+1)^{2n}}$$

and

$$E_{2n}(x) = \frac{(-1)^n 4(2n)!}{\pi^{2n+1}} \sum_{k=0}^{\infty} \frac{\sin(2k+1)\pi x}{(2k+1)^{2n+1}}.$$

So if we begin with the identity

$$\zeta(ps;x)\eta(qs;x) = \sum_{n_1=0}^{\infty}\sum_{n_2=0}^{\infty}(-1)^{n_2}\left[(n_1+n_2+x)^p(n_2+x)^q\right]^{-s}$$

$$-\sum_{n_1=0}^{\infty}\sum_{n_2=0}^{\infty}(-1)^{n_1+n_2}\left[(n_1+x)^p(n_1+x+n_2+1)^q\right]^{-s},$$

we get for all positive integers m and n,

$$B_{2m}(x)E_{2n-1}(x) = \sum_{r=0}^{m}\binom{2m}{2r}B_{2r}E_{2m+2n-2r-1}(x)$$

$$-m\sum_{r=1}^{n}\binom{2n-1}{2r-1}E_{2r-1}(0)E_{2m+2n-2r-1}(x)$$

and

$$B_{2m+1}(x)E_{2n}(x) = \sum_{r=0}^{m}\binom{2m+1}{2r}B_{2r}E_{2m+2n-2r+1}(x)$$

$$-\frac{2m+1}{2}\sum_{r=1}^{n}\binom{2n}{2r-1}E_{2r-1}(0)E_{2m+2n-2r+1}(x).$$

The following propositions then lead to the evaluations of $G^{--}_{2m,2n}$ and $G^{--}_{2m+1,2n+1}$.

Proposition 6.2.2. *For all positive integers m and n, we have*

$$\int_0^1 B_{2m}(x)E_{2n-1}(x)\sec(\pi x)\,dx = \frac{(-1)^{m+n}8(2m)!(2n-1)!}{2^{2m}\pi^{2m+2n}}G^{--}_{2m,2n}.$$

Proposition 6.2.3. *For each positive integer k, we have*

$$\int_0^1 E_{2k-1}(x)\sec(\pi x)\,dx = \frac{(-1)^k4(2k-1)!}{\pi^{2k}}L(2k).$$

Proposition 6.2.4. *For all positive integers m and n, we have*

$$\int_0^1 B_{2m+1}(x)E_{2n}(x)\sec(\pi x)\,dx$$

$$= \frac{(-1)^{m+n}8(2m+1)!(2n)!}{2^{2m+1}\pi^{2m+2n+2}}\left\{G^{--}_{2n+1,2n+1} - L(2n+1)\eta(2m+1)\right\}.$$

The evaluations of $G^{+-}_{2m,2n}$ and $G^{+-}_{2m+1,2n+1}$ depend on Bernoulli identities

$$B_{2m}(x)E_{2n-1}\left(x+\frac{1}{2}\right) = \sum_{r=0}^{m}\binom{2m}{2r}B_{2r}\left(\frac{1}{2}\right)E_{2m+2n-2r-1}\left(x+\frac{1}{2}\right)$$

$$-m\sum_{r=0}^{n-1}\binom{2n-1}{2r}E_{2r}\left(\frac{1}{2}\right)E_{2m+2n-2r-2}(x)$$

and

$$B_{2m+1}(x)E_{2n}\left(x+\frac{1}{2}\right) = \sum_{r=0}^{m}\binom{2m+1}{2r}B_{2r}\left(\frac{1}{2}\right)E_{2m+2n-2r+1}\left(x+\frac{1}{2}\right)$$

$$-\frac{m+1}{2}\sum_{r=0}^{n}\binom{2n}{2r}E_{2r}\left(\frac{1}{2}\right)E_{2m+2n-2r}(x).$$

Remark 6.2.5. As

$$\eta(s;x) = \sum_{n=0}^{\infty}(-1)^n(n+x)^{-s}$$

$$= \sum_{n=0}^{\infty}(2n+x)^{-s} - \sum_{n=0}^{\infty}(2n+1+x)^{-s}$$

$$= 2^{-s}\zeta\left(s;\frac{x}{2}\right) - 2^{-s}\zeta\left(s;\frac{x+1}{2}\right),$$

it is easy to see that

$$\eta(-m;x) = \frac{E_m(x)}{2} = \frac{2^m}{m+1}\left\{B_{m+1}\left(\frac{x+1}{2}\right) - B_{m+1}\left(\frac{x}{2}\right)\right\}.$$

Such a relation also follows from

$$\frac{2e^{xt}}{e^t+1} = \frac{2(e^{(x+1)t} - e^{xt})}{e^{2t}-1}.$$

6.2.3. Euler sums over rationally deformed simplices [28]

For a pair of positive integers p and q with $q \geq 2$ and any positive rational number $r = a/b$, we define

$$E^{(r)}_{p,q} = \sum_{k=1}^{\infty}\frac{1}{k^q}\sum_{j=1}^{[kr]}\frac{1}{j^p}.$$

Proposition 6.2.6. *For all positive integers* m, n *and a rational number* $r = a/b$ *with relatively prime positive integers* a, b, *we have*

$$\int_0^1 B_{2m}(\{bx\})B_{2n+1}(\{ax\})\cot(\pi x)\,dx$$
$$= \frac{(-1)^{m+n}4(2m)!(2n+1)!}{(2\pi)^{2m+2n+1}}\left\{E_{2m,2n+1}^{(r)} - \frac{1}{2a^{2m}b^{2n+1}}\zeta(2m+2n+1)\right\}.$$

The process of evaluating $E_{p,q}^{(r)}$ is similar to that of $E_{p,q}^{(k)}$. However, it is more difficult to carry out the evaluations.

6.3. Euler Sums on Hurwitz Zeta Functions

For a pair of positive integers p and q with $q \geq 2$ and positive real numbers x and y with x, $y \leq 1$, we define

$$H_{p,q}(x,y) = \sum_{k=0}^{\infty} \frac{1}{(k+y)^q} \sum_{j=0}^{\tilde{k}} \frac{1}{(j+x)^p}, \tag{6.3.1}$$

where

$$\tilde{k} = \begin{cases} k, & \text{if } x \leq y; \\ k-1, & \text{if } x > y. \end{cases}$$

The following reflection formulæ are immediate: For p, $q \geq 2$,

$$H_{p,q}(x,y) + H_{q,p}(y,x) = \zeta(p;x)\zeta(q;y) \tag{6.3.2}$$

if $x \neq y$, and

$$H_{p,q}(x,x) + H_{q,p}(x,x) = \zeta(p;x)\zeta(q;x) + \zeta(p+q;x). \tag{6.3.3}$$

Such new Euler sums provide a continuous version of, and actually building blocks for, the classical Euler sums or various kinds of generalized Euler sums.

Example 6.3.1. For the classical Euler sums, we have

$$H_{p,q}(1,1) = S_{p,q} \tag{6.3.4}$$

and for any positive integer N,

$$S_{p,q} = \frac{1}{N^{p+q}} \sum_{a=1}^{N} \sum_{b=1}^{N} H_{p,q} \left(\frac{a}{N}, \frac{b}{N} \right). \tag{6.3.5}$$

Example 6.3.2. Let $r = a/b$ with a and b being relatively prime positive integers. Then the extended Euler sum considered in [13],

$$E_{p,q}^{(r)} = \sum_{k=1}^{\infty} \frac{1}{k^q} \sum_{j=1}^{[kr]} \frac{1}{j^p}, \tag{6.3.6}$$

can be written as

$$E_{p,q}^{(r)} = \frac{1}{a^p b^q N^{p+q}} \sum_{u=1}^{aN} \sum_{v=1}^{bN} H_{p,q} \left(\frac{u}{aN}, \frac{v}{bN} \right). \tag{6.3.7}$$

Example 6.3.3. For Euler sums with Dirichlet character χ modulo N, as considered in [24],

$$S_{p,q}^{\chi} = \sum_{k=1}^{\infty} \frac{\chi(k)}{k^q} \sum_{j=1}^{k} \frac{1}{j^p}, \tag{6.3.8}$$

we note that

$$S_{p,q}^{\chi} = \frac{1}{N^{p+q}} \sum_{a=1}^{N} \sum_{b=1}^{N-1} \chi(b) H_{p,q} \left(\frac{a}{N}, \frac{b}{N} \right). \tag{6.3.9}$$

We may rewrite $S_{p,q}^{\chi}$ as

$$S_{p,q}^{\chi} = \frac{1}{2N^{p+q}} \sum_{b=1}^{N-1} \chi(b) \sum_{a=1}^{N-1} \left\{ H_{p,q} \left(\frac{a}{N}, \frac{b}{N} \right) + \chi(-1) H_{p,q} \left(\frac{N-a}{N}, \frac{N-b}{N} \right) \right\}$$
$$+ \frac{1}{2N^{p+q}} \sum_{b=1}^{N-1} \chi(b) \left\{ H_{p,q} \left(1, \frac{b}{N} \right) + \chi(-1) H_{p,q} \left(1, \frac{N-b}{N} \right) \right\}. \tag{6.3.10}$$

Consequently, $S_{p,q}^{\chi}$ can be explicitly evaluated provided that the quantities in the braces of (6.3.10) can be evaluated, which is the case if $\chi(-1) = (-1)^{p+q-1}$. This was proved, without giving explicit evaluations, in [24].

This last example motivates us to consider

$$T_{p,q}(x) = H_{p,q}(1,x) + \epsilon H_{p,q}(1, 1-x) \tag{6.3.11}$$

and

$$G_{p,q}(x,y) = H_{p,q}(x,y) + \epsilon H_{p,q}(1-x, 1-y),\qquad (6.3.12)$$

where

$$\epsilon = (-1)^{w-1} = (-1)^{p+q-1}\qquad (6.3.13)$$

so that $\epsilon = 1$ if the weight $w = p+q$ is odd and $\epsilon = -1$ when w is even.

The purpose of this section is exactly to give explicit evaluations of $T_{p,q}(x)$ and $G_{p,q}(x,y)$. It will be easier to work with their companions, which we now define. For $0 < x < 1$ and a pair of positive integers p and q with $q \geq 2$, define

$$E_{p,q}(x) = H_{p,q}(x,1) + \epsilon H_{p,q}(1-x, 1).\qquad (6.3.14)$$

In view of the reflection formula (which follows from (6.3.2))

$$E_{p,q}(x) + T_{q,p}(x) = \zeta(p;x)\zeta(q) + \epsilon\zeta(p; 1-x)\zeta(q)\qquad (6.3.15)$$

and the differentiation formula

$$E_{p,q}(x) = \frac{(-1)^{p-1}}{(p-1)!}\left(\frac{d}{dx}\right)^{p-1} E_{1,q}(x).\qquad (6.3.16)$$

For our convenience, we employ the notation for the logarithmic derivative of the gamma function:

$$\psi(x) = \frac{\Gamma'(x)}{\Gamma(x)}.$$

We shall start with obtaining explicit evaluation of various combinations of $H_{1,q}(x,y)$. We need the Kronecker limit formula for the Hurwitz zeta function,

$$\lim_{s\to 1+}\left(\zeta(s;x) - \frac{1}{s-1}\right) = -\psi(x),\qquad (6.3.17)$$

and its consequence

$$\sum_{j=0}^{\infty}\left(\frac{1}{j+x} - \frac{1}{j+y}\right) = -\psi(x) + \psi(y).\qquad (6.3.18)$$

Also, we have

$$\psi(x) - \psi(1-x) = -\pi\cot(\pi x).\qquad (6.3.19)$$

We shall also make use of the following partial fraction decomposition:

$$
\begin{aligned}
\frac{1}{(X+\alpha)^q X^p} &= \sum_{\ell=2}^{p} (-1)^{p+\ell} \binom{w-\ell-1}{q-1} \frac{1}{\alpha^{w-\ell} X^\ell} \\
&\quad + (-1)^p \sum_{\ell=2}^{q} \binom{w-\ell-1}{p-1} \frac{1}{\alpha^{w-\ell}(X+\alpha)^\ell} \\
&\quad + (-1)^{p+1} \binom{w-2}{q-1} \frac{1}{\alpha^{w-1}} \left(\frac{1}{X} - \frac{1}{X+\alpha} \right).
\end{aligned}
\tag{6.3.20}
$$

Proposition 6.3.4. *For $0 < x < 1$ and a positive integer q with $q \geq 2$, we have*

$$
\begin{aligned}
H_{1,q}(1,x) &= \frac{q}{2}\zeta(q+1;x) + \zeta(q;x)\{\psi(x)+\gamma\} \\
&\quad - \frac{1}{2}\sum_{\ell=2}^{q-1}\zeta(\ell;x)\zeta(q+1-\ell;x),
\end{aligned}
\tag{6.3.21}
$$

where γ is Euler's constant defined by

$$
\gamma = \lim_{n\to\infty} \left(\sum_{k=1}^{n} \frac{1}{k} - \log n \right).
$$

Proof. By definition, we have

$$
\begin{aligned}
H_{1,q}(1,x) &= \sum_{k=0}^{\infty} \frac{1}{(k+x)^q} \sum_{j=0}^{k-1} \frac{1}{j+1} \\
&= \sum_{k=0}^{\infty}\sum_{j=0}^{\infty} \frac{1}{(k+j+1+x)^q(j+1)}.
\end{aligned}
$$

In the partial fraction decomposition (6.3.20), let $p = 1$, then

$$
\frac{1}{(X+\alpha)^q X} = \frac{1}{\alpha^q}\left(\frac{1}{X} - \frac{1}{X+\alpha} \right) - \sum_{\ell=1}^{q-1} \frac{1}{\alpha^\ell(X+\alpha)^{q+1-\ell}}.
\tag{6.3.22}
$$

Using this we rewrite $H_{1,q}(1,x)$ as

$$
\begin{aligned}
&\sum_{k=0}^{\infty} \frac{1}{(k+x)^q} \sum_{j=0}^{\infty} \left(\frac{1}{j+1} - \frac{1}{k+j+1+x} \right) \\
&- \sum_{\ell=1}^{q-1}\sum_{k=0}^{\infty}\sum_{j=0}^{\infty} \frac{1}{(k+x)^\ell(k+j+1+x)^{q+1-\ell}},
\end{aligned}
$$

which, in view of (6.3.18) and the fact that $\psi(1) = -\gamma$, is equal to

$$\zeta(q; x) \{\gamma + \psi(x)\} + H_{1,q}(x, x) + (q - 1)\zeta(q + 1; x) - \sum_{\ell=1}^{q-1} H_{\ell, q+1-\ell}(x, x).$$

Our assertion then follows from the reflection formula (6.3.3).   □

The following propositions can be derived in a similar manner, and so we omit the proof.

Proposition 6.3.5. *For $0 < x < 1$ and a positive integer $q \geq 2$, we have*

$$H_{1,q}(x, x) + H_{1,q}(1 - x, 1) = \zeta(q + 1; x) - \zeta(q; x) \{\psi(1 - x) + \gamma\}$$

$$- \sum_{\ell=2}^{q-1} \zeta(\ell)\zeta(q + 1 - \ell; x).$$

$$(6.3.23)$$

Proposition 6.3.6. *For $0 < x < y < 1$ and a positive integer $q \geq 2$, we have*

$$H_{1,q}(x, y) + H_{1,q}(1 - x, y - x)$$
$$= \zeta(q; y - x) \{-\psi(x) + \psi(y)\} + \zeta(q; y) \{-\psi(1 - x) + \psi(y - x)\}$$
$$- \sum_{\ell=2}^{q-1} \zeta(\ell; y)\zeta(q + 1 - \ell; y - x).$$

$$(6.3.24)$$

For $0 < x < 1$ and a pair of positive integers p and q with $q \geq 2$, we recall

$$E_{p,q}(x) = H_{p,q}(x, 1) + \epsilon H_{p,q}(1 - x, 1),$$
$$T_{p,q}(x) = H_{p,q}(1, x) + \epsilon H_{p,q}(1, 1 - x)$$

and define

$$D_{p,q}(x) = H_{p,q}(1 - x, 1 - x) + \epsilon H_{p,q}(x, x).$$

Proposition 6.3.7. *For $0 < x < 1$ and each positive integer q with $q \geq 2$, let $\epsilon = (-1)^q$, then we have*

$$E_{1,q}(x) + D_{1,q}(x)$$
$$= \zeta(q + 1; 1 - x) + \epsilon \zeta(q + 1; x)$$
$$- \zeta(q; 1 - x) \{\psi(x) + \gamma\} - \epsilon \zeta(q; x) \{\psi(1 - x) + \gamma\}$$
$$- \sum_{\ell=2}^{q-1} \zeta(\ell) \{\zeta(q + 1 - \ell; 1 - x) + \epsilon \zeta(q + 1 - \ell; x)\}.$$

$$(6.3.25)$$

Proof. This follows from

$$E_{1,q}(x) + D_{1,q}(x) = \epsilon \{H_{1,q}(x,x) + H_{1,q}(1-x,1)\}$$
$$+ \{H_{1,q}(1-x,1-x) + H_{1,q}(x,1)\}$$

and the evaluation obtained in Proposition 6.3.5.   □

Proposition 6.3.8. *For $0 < x < 1$ and each positive integer q with $q \geq 2$, let $\epsilon = (-1)^q$, then*

$$E_{1,q}(x) - D_{1,q}(x)$$
$$= \zeta(q; 1-x) \{\psi(1-x) + \gamma\} + \epsilon\zeta(q; x) \{\psi(x) + \gamma\}$$
$$+ \sum_{\ell=2}^{q-1} (-1)^{\ell+1}\zeta(\ell) \{\zeta(q+1-\ell; 1-x) + \epsilon\zeta(q+1-\ell; x)\}. \tag{6.3.26}$$

Proof. To ease the burden of carrying around the ϵ factor in our derivation, we shall deal only with the case $q = 2n$ throughout, the case of odd q being similar.

Specializing the parameters $(p, q) = (2n - 1, 2)$ in (6.3.25), we get

$$\frac{1}{(X+\alpha)^2 X^{2n-1}} = \sum_{\ell=2}^{2n-1} (-1)^{\ell+1}(2n-\ell)\frac{1}{\alpha^{2n+1-\ell}X^\ell} - \frac{1}{\alpha^{2n+1-2}(X+\alpha)^2}$$
$$+ (2n-1)\frac{1}{\alpha^{2n}}\left(\frac{1}{X} - \frac{1}{X+\alpha}\right).$$

It follows that

$$H_{2n-1,2}(x,x) = \zeta(2n+1; x) + \sum_{\ell=2}^{2n-1} (-1)^{\ell+1}(2n-\ell)\zeta(2n+1-\ell)\zeta(\ell; x)$$
$$- H_{2n-1,2}(1, x) + (2n-1)H_{1,2n}(x, 1).$$

Replacing x by $1 - x$ and adding together, we get

$$T_{2n-1,2}(x) + D_{2n-1,2}(x)$$
$$= \zeta(2n+1; x) + \zeta(2n+1; 1-x) + (2n-1)E_{1,2n}(x)$$
$$+ \sum_{\ell=2}^{2n-1} (-1)^{\ell+1}(2n-\ell)\zeta(2n+1-\ell) \{\zeta(\ell; x) + \zeta(\ell; 1-x)\}. \tag{6.3.27}$$

In the same way, we get

$$
\begin{aligned}
&T_{2n-1,2}(x) + D_{2n-1,2}(x) \\
&= \zeta(2n+1;x) + \zeta(2n+1;1-x) + (2n-1)\zeta(2n;x)\{\gamma+\psi(x)\} \\
&\quad + (2n-1)\zeta(2n;1-x)\{\gamma+\psi(1-x)\} + (2n-1)D_{1,2n}(x) \\
&\quad + \sum_{\ell=2}^{2n-1}(-1)^{\ell+1}(2n-\ell)\{\zeta(2n+1-\ell;x) + \zeta(2n+1-\ell;1-x)\}\zeta(\ell)
\end{aligned}
$$

$$(6.3.28)$$

if we begin with

$$
H_{2n-1,2}(1,x) = \sum_{k=0}^{\infty}\sum_{j=0}^{\infty}\frac{1}{(k+j+1+x)^2(j+1)^{2n-1}}.
$$

Subtraction of (6.3.27) from (6.3.28) then yields the result.   □

Theorem 6.3.9. *Let $E_{p,q}(x)$ be defined by (6.3.14). For each positive integer $q \geq 2$, let $w = 1+q$ and $\epsilon = (-1)^q$, then*

$$
\begin{aligned}
E_{1,q}(x) &= \frac{1}{2}\{\zeta(w-1;1-x) - \epsilon\zeta(w-1;x)\}\pi\cot(\pi x) \\
&\quad - \sum_{\ell=0}^{[(q-1)/2]}\zeta(2\ell)\{\zeta(w-2\ell;1-x) + \epsilon\zeta(w-2\ell;x)\},
\end{aligned}
$$

$$(6.3.29)$$

where, as customary, $\zeta(0) = -1/2$.

Corollary 6.3.10. *For all positive integers p and q with $q \geq 2$, let $w = p+q$ and $\epsilon = (-1)^{p+q-1}$, then we have*

$$
\begin{aligned}
E_{p,q}(x) &= (-1)^{p-1}\Bigg\{\frac{1}{2}\binom{w-2}{q-1}\{\zeta(w-1;1-x) - \epsilon\zeta(w-1;x)\}\pi\cot(\pi x) \\
&\quad - \frac{1}{2}\sum_{\ell=1}^{p-1}\binom{w-\ell-2}{q-1}\{\zeta(\ell+1;1-x) + (-1)^{\ell+1}\zeta(\ell+1;x)\} \\
&\qquad \times \{\zeta(w-\ell-1;1-x) + \epsilon(-1)^{\ell+1}\zeta(w-\ell-1;x)\} \\
&\quad - \sum_{\ell=0}^{[(q-1)/2]}\binom{w-2\ell-1}{p-1}\zeta(2\ell)\{\zeta(w-2\ell;1-x) + \epsilon\zeta(w-2\ell;x)\}\Bigg\}.
\end{aligned}
$$

$$(6.3.30)$$

Proofs of Theorem 6.3.9 and Corollary 6.3.10. The evaluation of $E_{1,q}(x)$ follows from Propositions 6.3.7 and 6.3.8. Now Corollary 6.3.10 is an easy consequence in virtue of (6.3.19) and the facts that

$$\frac{d}{dx}\zeta(s;1-x) = s\zeta(s+1;1-x),$$

$$\frac{d}{dx}\zeta(s;x) = -s\zeta(s+1;x)$$

and

$$\frac{d}{dx}(\psi(x) - \psi(1-x)) = \zeta(2;x) + \zeta(2;1-x). \qquad \square$$

Remark 6.3.11. Note that $D_{p,q}(x)$ can be expressed in terms of $E_{u,v}(x)$ and $T_{u,v}(x)$ with the same weight $u + v = p + q$. Besides, just like the differentiation formula (6.3.16), the higher derivatives of $D_{1,q}(x)$ also give recursively the evaluation of $D_{p,q}(x)$.

Recall that for a pair of positive integers p and q with $q \geq 2$, and $0 < x < y < 1$,

$$G_{p,q}(x,y) = H_{p,q}(x,y) + \epsilon H_{p,q}(1-x, 1-y)$$

with $\epsilon = (-1)^{p+q-1}$.

In a similar manner, we obtain the following theorem.

Theorem 6.3.12. *For $0 < x < y < 1$ and each positive integer n, we have*

$$2G_{1,2n}(x,y)$$
$$= \{\zeta(2n; y - x) - \zeta(2n; 1 + x - y)\}\{\pi \cot(\pi x) - \pi \cot(\pi y)\}$$
$$+ \{\zeta(2n; x) + \zeta(2n; 1 - x)\}\{-\psi(y) - \psi(1-y) + \psi(y-x) + \psi(1+x-y)\}$$
$$- \sum_{\ell=2}^{2n-1}\{\zeta(2n+1-\ell; y-x) + (-1)^{\ell}\zeta(2n+1-\ell; 1+x-y)\}$$
$$\times \{\zeta(\ell; 1-x) + (-1)^{\ell}\zeta(\ell; x)\}.$$

$$(6.3.31)$$

Remark 6.3.13. Since we have the following differentiation formulæ,

$$G_{p,q}(x,y) = \frac{(-1)^{p-1}}{(p-1)!}\left(\frac{\partial}{\partial x}\right)^{p-1} G_{1,q}(x,y)$$

and

$$G_{p,q+1}(x,y) = \frac{-1}{q}\left(\frac{\partial}{\partial y}\right)G_{p,q}(x,y),$$

it is not difficult to derive explicit evaluation of $H_{p,q}(x,y) + H_{p,q}(1-x,1-y)$ for odd w and of $H_{p,q}(x,y) - H_{p,q}(1-x,1-y)$ for even w from Theorem 6.3.12 if one so wishes.

6.4. Multiple Zeta Values

Multiple zeta values are natural generalizations of the classical Euler sums. For positive integers $\alpha_1, \alpha_2, \ldots, \alpha_r$ with $\alpha_r \geq 2$, the multiple zeta value or r-fold Euler sum is defined as

$$\zeta(\alpha_1, \alpha_2, \ldots, \alpha_r) = \sum_{1 \leq k_1 < k_2 < \cdots < k_r} k_1^{-\alpha_1} k_2^{-\alpha_2} \cdots k_r^{-\alpha_r}. \qquad (6.4.1)$$

By changing of dummy variables, the multiple series is equal to

$$\sum_{k_1=1}^{\infty} \sum_{k_2=1}^{\infty} \cdots \sum_{k_r=1}^{\infty} k_1^{-\alpha_1}(k_1+k_2)^{-\alpha_2} \cdots (k_1+k_2+\cdots+k_r)^{-\alpha_r} \qquad (6.4.2)$$

or in nested form as

$$\sum_{k_r=1}^{\infty} \frac{1}{k_r^{\alpha_r}} \sum_{k_{r-1}=1}^{k_r-1} \frac{1}{k_{r-1}^{\alpha_{r-1}}} \cdots \sum_{k_1=1}^{k_2-1} \frac{1}{k_1^{\alpha_1}}. \qquad (6.4.3)$$

Around 1994, C. Markett [41] and later J. M. Borwein and R. Girgensohn [10] obtained, among other things, that for $n \geq 3$,

$$\zeta(1,1,n) = \frac{n(n+1)}{6}\zeta(n+2) - \frac{n-1}{2}\zeta(2)\zeta(n)$$

$$- \frac{n}{4}\sum_{k=0}^{n-4}\zeta(n-k-1)\zeta(k+3) \qquad (6.4.4)$$

$$+ \frac{1}{6}\sum_{k=0}^{n-4}\zeta(n-k-2)\sum_{j=0}^{k}\zeta(k-j+2)\zeta(j+2).$$

This is just the beginning of evaluating multiple zeta values with depth greater than 2. Here we begin with the evaluation of $\zeta(1,1,n)$ and then triple Euler sums of even weight.

Theorem 6.4.1. *For each positive integer n, we have*

$$\zeta(1,1,2n) = \zeta(2n+2) + \sum_{\alpha=2}^{2n-1} (-1)^{\alpha+1} \zeta(2n+1-\alpha)\zeta(1,\alpha) \qquad (6.4.5)$$

and

$$3\zeta(1,1,2n+1) = -\{\zeta(2n+3) + 2\zeta(1,2n+2) + 2\zeta(2,2n+1)\}$$
$$+ \sum_{\alpha=2}^{2n} (-1)^{\alpha} \zeta(2n+2-\alpha)\zeta(1,\alpha).$$

Proof. Consider the multiple zeta value

$$\zeta(1,2n-1,2) = \sum_{j=1}^{\infty}\sum_{k=1}^{\infty}\sum_{\ell=1}^{\infty} \frac{1}{j(k+j)^{2n-1}(k+j+\ell)^2}.$$

Employing the partial fraction decomposition

$$\frac{1}{X^{2n-1}(X+T)^2} = \sum_{\alpha=2}^{2n-1} (-1)^{\alpha+1}(2n-\alpha)\frac{1}{T^{2n+1-\alpha}X^\alpha} - \frac{1}{T^{2n-1}(X+T)^2}$$
$$+ (2n-1)\frac{1}{T^{2n}}\left(\frac{1}{X} - \frac{1}{X+T}\right)$$

$$(6.4.6)$$

with $X = k+j$ and $T = \ell$, we get

$$\zeta(1,2n-1,2) = \sum_{\alpha=2}^{2n-1} (-1)^{\alpha+1}(2n-\alpha)\zeta(2n+1-\alpha)\zeta(1,\alpha)$$
$$- \eta(1,2n-1,2)$$
$$+ (2n-1)\sum_{\ell=1}^{\infty}\frac{1}{\ell^{2n}}\sum_{j=1}^{\infty}\frac{1}{j}\sum_{k=1}^{\infty}\left(\frac{1}{k+j} - \frac{1}{k+j+\ell}\right),$$

$$(6.4.7)$$

where

$$\eta(1,2n-1,2) = \sum_{j=1}^{\infty}\sum_{k=1}^{\infty}\sum_{\ell=1}^{\infty} \frac{1}{j\ell^{2n-1}(j+k+\ell)^2}.$$

The triple series in $\zeta(1, 2n - 1, 2)$ is equal to

$$\sum_{\ell=1}^{\infty} \frac{1}{\ell^{2n}} \sum_{j=1}^{\infty} \frac{1}{j} \sum_{k=1}^{\ell} \frac{1}{k+j} = \sum_{\ell=1}^{\infty} \frac{1}{\ell^{2n}} \sum_{k=1}^{\ell} \frac{1}{k} \sum_{j=1}^{\infty} \left(\frac{1}{j} - \frac{1}{k+j} \right)$$

$$= \sum_{\ell=1}^{\infty} \frac{1}{\ell^{2n}} \sum_{k=1}^{\ell} \frac{1}{k} \sum_{j=1}^{k} \frac{1}{j}.$$

On the other hand, if we begin with $\eta(1, 2n - 1, 2)$ and employ the same partial fraction decomposition, we get

$$\eta(1, 2n - 1, 2) = \sum_{\alpha=2}^{2n-1} (-1)^{\alpha+1} (2n - \alpha) \zeta(\alpha) \zeta(1, 2n + 1 - \alpha)$$

$$- \zeta(1, 2n - 1, 2) \tag{6.4.8}$$

$$+ (2n - 1) \sum_{k=1}^{\infty} \frac{1}{k^{2n}} \sum_{j=1}^{k-1} \frac{1}{j} \sum_{\ell=1}^{k} \frac{1}{\ell}.$$

The difference of (6.4.7) and (6.4.8) gives

$$\sum_{k=1}^{\infty} \frac{1}{k^{2n}} \sum_{j=1}^{k-1} \frac{1}{j} \sum_{\ell=1}^{k} \frac{1}{\ell} - \sum_{\ell=1}^{\infty} \frac{1}{\ell^{2n}} \sum_{k=1}^{\ell} \frac{1}{k} \sum_{j=1}^{k} \frac{1}{j}$$

$$= \sum_{\alpha=2}^{2n-1} (-1)^{\alpha+1} \zeta(2n + 1 - \alpha) \zeta(1, \alpha). \tag{6.4.9}$$

The left-hand side of (6.4.9) is equal to

$$\{2\zeta(1, 1, 2n) + \zeta(1, 2n + 1) + \zeta(2, 2n)\}$$

$$- \{\zeta(1, 1, 2n) + \zeta(1, 2n + 1) + \zeta(2, 2n) + \zeta(2n + 2)\}$$

by an elementary consideration, so we get the evaluation of $\zeta(1, 1, 2n)$. To get the evaluation of $\zeta(1, 1, 2n + 1)$, we begin with the multiple zeta value $\zeta(1, 2n, 2)$ and get the identity

$$\sum_{k=1}^{\infty} \frac{1}{k^{2n+1}} \sum_{j=1}^{k-1} \frac{1}{j} \sum_{\ell=1}^{k} \frac{1}{\ell} + \sum_{\ell=1}^{\infty} \frac{1}{\ell^{2n+1}} \sum_{k=1}^{\ell} \frac{1}{k} \sum_{j=1}^{k} \frac{1}{j}$$

$$= \sum_{\alpha=2}^{2n} (-1)^{\alpha} \zeta(2n + 2 - \alpha) \zeta(1, \alpha). \tag{6.4.10}$$

Therefore, the evaluation of $\zeta(1, 1, 2n + 1)$ follows from (6.4.10). $\square$

Now we introduce two families of triple Euler sums on Hurwitz zeta function. For all positive integers p, q and r with $r \geq 2$ and positive real numbers x, y and z, we define

$$H_{p,q,r}(x,y,z) = \sum_{\ell=0}^{\infty} \frac{1}{(\ell+z)^r} \sum_{k=0}^{\tilde{\ell}} \frac{1}{(k+y)^q} \sum_{j=0}^{\tilde{k}} \frac{1}{(j+x)^p} \qquad (6.4.11)$$

and

$$G_{p,q,r}(x,y,z) = \sum_{\ell=0}^{\infty} \frac{1}{(\ell+z)^r} \sum_{k=0}^{\tilde{\ell}} \frac{1}{(k+y)^q} \sum_{j=0}^{\tilde{\ell}} \frac{1}{(j+x)^p}. \qquad (6.4.12)$$

Consider a triple series with two variables x and y as

$$H_{1,p,q}(x,y,y) - H_{1,p+q}(x,y)$$

$$= \sum_{j=0}^{\infty} \sum_{k=0}^{\infty} \sum_{\ell=1}^{\infty} \frac{1}{(j+x)(k+j+y)^p(k+j+\ell+y)^q}.$$

Again we focus on the cases when $p = 2n - 1$, $q = 2$ and $p = 2n$, $q = 2$ with $0 < x, y \leq 1$. Proceeding as before, we get the following theorem.

Theorem 6.4.2. *For each positive integer $n \geq 2$ and positive real numbers x, y with $0 < x < y \leq 1$, we have*

$$G_{1,1,2n}(y,x,y) - H_{1,1,2n}(y,y-x,1)$$
$$= H_{1,2n}(y-x,1)\left\{-\psi(x) + \psi(y)\right\} - H_{1,2n}(x,y)\left\{\gamma + \psi(y)\right\}$$
$$+ \sum_{\alpha=2}^{2n-1} (-1)^{\alpha+1}\zeta(2n+1-\alpha)H_{1,\alpha}(x,y) \qquad (6.4.13)$$

and

$$G_{1,1,2n+1}(y,x,y) + H_{1,1,2n+1}(y,y-x,1)$$
$$= H_{1,2n+1}(y-x,1)\left\{\psi(x) - \psi(y)\right\} - H_{1,2n+1}(x,y)\left\{\gamma + \psi(y)\right\}$$
$$+ \sum_{\alpha=2}^{2n}(-1)^{\alpha}\zeta(2n+2-\alpha)H_{1,\alpha}(x,y). \qquad (6.4.14)$$

Here we demonstrate the general procedure to obtain the evaluation of $\zeta(p,q,r)$ when the weight $p+q+r$ is even from (6.4.13) and (6.4.14). To obtain the values of $\zeta(1, 2, 2n + 1)$ and $\zeta(2, 1, 2n + 1)$, we apply the partial differential operators $-\partial/\partial x$ and $-\partial/\partial y$ to both sides of (6.4.14) and then set $x = y = 1$ to get

$$\sum_{k=1}^{\infty} \frac{1}{k^{2n+1}} \sum_{j=1}^{k} \frac{1}{j^2} \sum_{\ell=1}^{k} \frac{1}{\ell} - \sum_{k=1}^{\infty} \frac{1}{k^{2n+1}} \sum_{j=1}^{k-1} \frac{1}{j^2} \sum_{\ell=1}^{j} \frac{1}{\ell}$$

$$= -\zeta(2)\zeta(1, 2n + 1) - \zeta(3)\zeta(2n + 1) \qquad (6.4.15)$$

$$+ \sum_{\alpha=2}^{2n} (-1)^{\alpha} \zeta(2n + 2 - \alpha) S_{2,\alpha}$$

and

$$(2n + 1) \sum_{k=1}^{\infty} \frac{1}{k^{2n+2}} \sum_{j=1}^{k} \frac{1}{j} \sum_{\ell=1}^{k} \frac{1}{\ell} + \sum_{k=1}^{\infty} \frac{1}{k^{2n+1}} \sum_{j=1}^{k} \frac{1}{j} \sum_{\ell=1}^{k} \frac{1}{\ell^2}$$

$$+ \sum_{k=1}^{\infty} \frac{1}{k^{2n+1}} \sum_{j=1}^{k-1} \frac{1}{j^2} \sum_{\ell=1}^{j} \frac{1}{\ell} + \sum_{k=1}^{\infty} \frac{1}{k^{2n+1}} \sum_{j=1}^{k-1} \frac{1}{j} \sum_{\ell=1}^{j} \frac{1}{\ell^2} \qquad (6.4.16)$$

$$= \zeta(2)\zeta(1, 2n + 1) - \zeta(3)\zeta(2n + 1) + \zeta(2)S_{1,2n+1}$$

$$+ \sum_{\alpha=2}^{2n} (-1)^{\alpha} \alpha \zeta(2n + 2 - \alpha) S_{1,\alpha+1}.$$

From (6.4.15), we get the evaluation of $\zeta(2, 1, 2n + 1)$ and from (6.4.16), we get the evaluation of $2\zeta(2, 1, 2n + 1) + 2\zeta(1, 2, 2n + 1)$.

Once we obtain the values of $\zeta(2, 1, 2n + 1)$ and $\zeta(1, 2, 2n + 1)$, we proceed to evaluate $\zeta(1, 3, 2n)$, $\zeta(3, 1, 2n)$ and $\zeta(2, 2, 2n)$. Applying the partial differential operators

$$\frac{\partial^2}{\partial x^2}, \quad \frac{\partial^2}{\partial x \partial y}, \quad \frac{\partial^2}{\partial y^2}$$

to both sides of (6.4.13) and then setting $x = y = 1$, we get a system of three linear equations in $\zeta(1, 3, 2n)$, $\zeta(3, 1, 2n)$ and $\zeta(2, 2, 2n)$. So our evaluation follows by solving the linear system.

For our convenience, we let $\{1\}^m$ be m repetitions of 1. For example,

$$\zeta(\{1\}^3, n) = \zeta(1, 1, 1, n) \quad \text{and} \quad \zeta(\{1\}^5, m) = \zeta(1, 1, 1, 1, 1, m).$$

Multiple zeta values of the form $\zeta(\{1\}^m, n)$ worth a special attention. For all positive integers m and n, we have

$$\zeta(\{1\}^{m-1}, n + 1) = \zeta(\{1\}^{n-1}, m + 1). \qquad (6.4.17)$$

This is referred to *Drinfeld duality theorem* and it can be proved if we express it as

$$\zeta(\{1\}^{m-1}, n+1) = \int_{0<t_1<t_2<\cdots<t_{m+n}<1} \prod_{j=1}^{m} \frac{dt_j}{1-t_j} \prod_{k=m+1}^{m+n} \frac{dt_k}{t_k}. \quad (6.4.18)$$

Also there is a generating function for $\zeta(\{1\}^m, n+2)$ [9, 11] given by

$$\sum_{m=0}^{\infty} \sum_{n=0}^{\infty} x^{n+1} y^{m+1} \zeta(\{1\}^m, n+2)$$

$$= 1 - \frac{\Gamma(1-x)\Gamma(1-y)}{\Gamma(1-x-y)} \quad (6.4.19)$$

$$= 1 - \exp\left\{\sum_{k=2}^{\infty} (x^k + y^k - (x+y)^k) \frac{\zeta(k)}{k}\right\}.$$

Therefore, we are able to express $\zeta(\{1\}^m, n)$ in terms of special values at positive integers of Riemann zeta function. To do so, we need to introduce Euler sums with two branches.

For all positive integers p, q and n with $n \geq 2$, we define a *Euler sum with two branches* as

$$G_n(p,q) = \sum_{k_0=1}^{\infty} \frac{1}{k_0^n} \sum_{k_1=1}^{k_0} \frac{1}{k_1} \cdots \sum_{k_q=1}^{k_{q-1}} \frac{1}{k_q} \left(\sum_{\ell_1=1}^{k_0-1} \frac{1}{\ell_1} \sum_{\ell_2=1}^{\ell_1-1} \frac{1}{\ell_2} \cdots \sum_{\ell_p=1}^{\ell_{p-1}-1} \frac{1}{\ell_p}\right). \quad (6.4.20)$$

When $p = 0$, this is just a sum of multiple zeta values [44]

$$G_n(0,q) = \sum_{r=1}^{q+1} \sum_{|\alpha|=q+1} \zeta(\alpha_1, \alpha_2, \ldots, \alpha_r + n - 1). \quad (6.4.21)$$

Therefore, for general q, we have the following decomposition

$$G_n(p,q) = \sum_{r=p+1}^{p+q+1} \binom{r-1}{p} \sum_{|\alpha|=p+q+1} \zeta(\alpha_1, \alpha_2, \ldots, \alpha_r + n - 1). \quad (6.4.22)$$

Proposition 6.4.3. *For all positive integers m, n and p with $m \geq 2$ and $1 \leq p \leq m/2$, we have*

$$G_{2n}(p, m-p) - G_{2n}(m-1-p, p+1)$$

$$= \sum_{\alpha=2}^{2n-1} (-1)^{\alpha+1} \zeta(\{1\}^p, 2n+1-\alpha) \zeta(\{1\}^{m-p-1}, \alpha) \quad (6.4.23)$$

and

$$G_{2n+1}(p, m-p) + G_{2n+1}(m-1-p, p+1)$$

$$= \sum_{\alpha=2}^{2n} (-1)^\alpha \zeta(\{1\}^p, 2n+2-\alpha) \zeta(\{1\}^{m-p-1}, \alpha). \quad (6.4.24)$$

Proof. Here we only prove the case when $m = 5$ and $p = 3$. The general cases follow with a similar procedure. For $\mathbf{u} = (1, 1, 1, 2n-1, 1, 2)$, we consider the multiple series

$$\eta(\mathbf{u}) = \sum_{\mathbf{k} \in \mathbb{N}^6} \frac{1}{s_1 s_2 s_3 s_4^{2n-1} k_5 s_6^2}, \quad (6.4.25)$$

where $s_j = k_1 + k_2 + \cdots + k_j$, $j = 1, 2, 3, 4, 5, 6$. Employing the partial fraction decomposition

$$\frac{1}{X^{2n-1}(X+T)^2} = \sum_{\alpha=2}^{2n-1} (-1)^{\alpha+1}(2n-\alpha) \frac{1}{T^{2n+1-\alpha} X^\alpha}$$

$$- \frac{1}{T^{2n-1}(X+T)^2} + \frac{2n-1}{T^{2n}} \left(\frac{1}{X} - \frac{1}{X+T} \right) \quad (6.4.26)$$

with $X = s_4$ and $T = k_5 + k_6 = s_6 - s_4$, we get

$$\eta(\mathbf{u}) = \sum_{\alpha=2}^{2n-1} (-1)^{\alpha+1}(2n-\alpha) \zeta(1, 2n+1-\alpha) \zeta(\{1\}^3, \alpha) - \widetilde{\eta}(\mathbf{u})$$

$$+ (2n-1) \sum_{k_5, k_6=1}^{\infty} \frac{1}{(k_5+k_6)^{2n} k_5} \sum_{k_1, k_2, k_3=1}^{\infty} \frac{1}{s_1 s_2 s_3} \sum_{k_4=1}^{\infty} \left(\frac{1}{s_4} - \frac{1}{s_6} \right),$$

where

$$\widetilde{\eta}(\mathbf{u}) = \sum_{\mathbf{k} \in \mathbb{N}^6} \frac{1}{s_1 s_2 s_3 (k_5+k_6)^{2n-1} k_5 s_6^2}.$$

Note that

$$\sum_{k_4=1}^{\infty} \left(\frac{1}{s_4} - \frac{1}{s_6} \right) = \sum_{k_4=1}^{k_5+k_6} \frac{1}{s_3+k_4}$$

and

$$\sum_{k_3=1}^{\infty} \sum_{k_4=1}^{k_5+k_6} \frac{1}{s_3(s_3+k_4)} = \sum_{k_4=1}^{k_5+k_6} \frac{1}{k_4} \sum_{k_3=1}^{\infty} \left(\frac{1}{k_1+k_2+k_3} - \frac{1}{k_1+k_2+k_3+k_4} \right)$$

$$= \sum_{k_4=1}^{k_5+k_6} \frac{1}{k_4} \sum_{k_3=1}^{k_4} \frac{1}{k_1+k_2+k_3}.$$

Repeating the same process for k_1 and k_2, we get

$$\eta(\mathbf{u}) = \sum_{\alpha=2}^{2n-1} (-1)^{\alpha+1}(2n-\alpha)\zeta(1,2n+1-\alpha)\zeta(\{1\}^3,\alpha)$$
$$- \widetilde{\eta}(\mathbf{u}) + (2n-1)G_{2n}(1,4). \tag{6.4.27}$$

Exchanging the roles of η and $\widetilde{\eta}$, we get

$$\widetilde{\eta}(\mathbf{u}) = \sum_{\alpha=2}^{2n-1} (-1)^{\alpha+1}(2n-\alpha)\zeta(1,\alpha)\zeta(\{1\}^3,2n+1-\alpha)$$
$$- \eta(\mathbf{u}) + (2n-1)G_{2n}(3,2). \tag{6.4.28}$$

The difference of (6.4.27) and (6.4.28) leads to

$$G_{2n}(3,2) - G_{2n}(1,4) = \sum_{\alpha=2}^{2n-1} (-1)^{\alpha+1}\zeta(1,2n+1-\alpha)\zeta(\{1\}^3,\alpha). \tag{6.4.29}$$

On the other hand, if we begin with the multiple series

$$\eta(\mathbf{v}) = \sum_{\mathbf{k}\in\mathbb{N}^6} \frac{1}{s_1 s_2 s_3 s_4^{2n} k_5 s_6^2}, \qquad \mathbf{v} = (1,1,1,2n,1,2), \tag{6.4.30}$$

and employ another partial fraction decomposition

$$\frac{1}{X^{2n}(X+T)^2} = \sum_{\alpha=2}^{2n} (-1)^{\alpha}(2n+1-\alpha)\frac{1}{T^{2n+2-\alpha}X^\alpha}$$
$$+ \frac{1}{T^{2n}(X+T)^2} - \frac{2n}{T^{2n+1}}\left(\frac{1}{X} - \frac{1}{X+T}\right), \tag{6.4.31}$$

we get

$$G_{2n+1}(3,2) + G_{2n+1}(1,4) = \sum_{\alpha=2}^{2n} (-1)^{\alpha}\zeta(1,2n+2-\alpha)\zeta(\{1\}^3,\alpha). \tag{6.4.32}$$

This proves the case of $m = 5$ and $p = 2$.   □

Remark 6.4.4. For $m = p + q$, we have a series expression for $G_n(p,q)$, $n \geq 3$, as

$$\sum_{\mathbf{k}\in\mathbb{N}^{p+1}} \sum_{\mathbf{j}\in\mathbb{N}^q} \frac{1}{s_1 s_2 \cdots s_p s_{p+1}^{n-1} \sigma_1 \sigma_2 \cdots \sigma_q(s_{p+1}+\sigma_q)},$$

where $s_m = k_1 + k_2 + \cdots + k_m$ and $\sigma_\ell = j_1 + j_2 + \cdots + j_\ell$.

For our convenience, we let for $1 \le r \le m$ that

$$S_r = \sum_{|\alpha|=m+1} \zeta(\alpha_1, \alpha_2, \ldots, \alpha_r + 2n - 1) \qquad (6.4.33)$$

and

$$\overline{S}_r = \sum_{|\alpha|=m+1} \zeta(\alpha_1, \alpha_2, \ldots, \alpha_r + 2n). \qquad (6.4.34)$$

We divide our cases into four parts according to even or odd of the positive integers m and n.

Part 1. The evaluation of $\zeta(\{1\}^{2k}, 2n)$

First we decompose $G_{2n}(p, q)$ with $p + q = 2k$ into linear combinations of $F = \zeta(\{1\}^{2k}, 2n)$ and S_j, $j = 1, 2, \ldots, 2k$ by (6.4.22). With $C_k^n = \binom{n}{k}$, the result is as follows:

$$
\begin{aligned}
G_{2n}(0, 2k) &= & F+ & & S_{2k}+ & & S_{2k-1} + \cdots + S_3 + S_2 + S_1, \\
G_{2n}(1, 2k-1) &= & C_1^{2k} F+ & & C_1^{2k-1} S_{2k}+ & & C_1^{2k-2} S_{2k-1} + \cdots + 2S_3 + S_2, \\
G_{2n}(2, 2k-2) &= & C_2^{2k} F+ & & C_2^{2k-1} S_{2k}+ & & C_2^{2k-2} S_{2k-1} + \cdots + S_3,
\end{aligned}
$$

$$\cdots\cdots\cdots\cdots\cdots\cdots\cdots\cdots\cdots\cdots\cdots\cdots\cdots$$

$$
\begin{aligned}
G_{2n}(2k-3, 3) &= C_{2k-3}^{2k} F+ & C_{2k-3}^{2k-1} S_{2k}+ & C_{2k-3}^{2k-2} S_{2k-1} + S_{2k-2}, \\
G_{2n}(2k-2, 2) &= C_{2k-2}^{2k} F+ & C_{2k-2}^{2k-1} S_{2k}+ & S_{2k-1}, \\
G_{2n}(2k-1, 1) &= C_{2k-1}^{2k} F+ & S_{2k}.
\end{aligned}
$$

$$(6.4.35)$$

On the other hand, by Proposition 6.4.3, we have for $0 \le p \le 2k - 1$ that

$$
\begin{aligned}
& G_{2n}(p, 2k - p) - G_{2n}(2k - 1 - p, p + 1) \\
&= \sum_{\alpha=2}^{2n-1} (-1)^{\alpha+1} \zeta(\{1\}^p, 2n + 1 - \alpha) \zeta(\{1\}^{2k-p-1}, \alpha).
\end{aligned}
\qquad (6.4.36)
$$

Thus the alternating sums

$$\sum_{p=0}^{2k-1} (-1)^{p+1} G_{2n}(p, 2k - p)$$

is our required combination to produce the evaluation of $\zeta(\{1\}^{2k}, 2n)$. It is equal to

$$\zeta(\{1\}^{2k}, 2n) - \zeta(2n + 2k)$$

by (6.4.35) and it is equal to

$$\sum_{p=0}^{k-1}\sum_{\alpha=2}^{2n-1}(-1)^{\alpha+p+1}\zeta(\{1\}^p,2n+1-\alpha)\zeta(\{1\}^{2k-p-1},\alpha)$$

by (6.4.36). Consequently, we get that

$$\zeta(\{1\}^{2k},2n)=\zeta(2n+2k)$$
$$+\sum_{p=0}^{k-1}\sum_{\alpha=2}^{2n-1}(-1)^{\alpha+p+1}\zeta(\{1\}^p,2n+1-\alpha)\zeta(\{1\}^{2k-p-1},\alpha).$$

Part 2. The evaluation of $\zeta(\{1\}^{2k+1},2n+1)$

Like our previous case, the alternating sum

$$\sum_{p=0}^{2k}(-1)^p G_{2n+1}(p,2k+1-p)$$

is equal to

$$\zeta(\{1\}^{2k+1},2n+1)+\zeta(2k+2n+2)$$

by the decompositions of $G_{2n+1}(p,q)$ for $p+q=2k+1$. Also it is equal to

$$\{G_{2n+1}(2k,1)+G_{2n+1}(0,2k+1)\}-\{G_{2n+1}(2k-1,2)+G_{2n+1}(1,2k)\}$$
$$+\cdots+(-1)^k G_{2n+1}(k+1,k).$$

This leads to the evaluation

$$\zeta(\{1\}^{2k+1},2n+1)=-\zeta(2k+2n+2)$$
$$+\frac{1}{2}\sum_{p=0}^{2k}\sum_{\alpha=2}^{2n}(-1)^{\alpha+p}\zeta(\{1\}^p,2n+2-\alpha)\zeta(\{1\}^{2k-p},\alpha).$$

Part 3. The evaluation of $\zeta(\{1\}^{2k},2n+1)$

Instead of the simple alternating sum

$$\sum_{p=1}^{m}(-1)^{p+1}G_{2n+1}(p,m-p)$$

which no longer works, we consider weighted alternating sum in an attempt to eliminate $\overline{S}_3,\overline{S}_4,\ldots,\overline{S}_{2k}$ in the combination.

We need the following simple fact about binomial coefficients.

Proposition 6.4.5. *For any positive integer $k \geq 2$, we define vectors* $\mathbf{u}$ *and* $\mathbf{v}_j$, $j = 2, 3, \ldots, 2k$ *in* $\mathbb{R}^{2k}$ *as*

$$\mathbf{u} = (2k - 1, -(2k - 3), \ldots, (-1)^k, (-1)^k, \ldots, -(2k - 3), 2k - 1),$$

$$\mathbf{v}_j = (C_0^j, C_1^j, \ldots, C_j^j, \{0\}^{2k-j-1}), \quad j = 2, 3, \ldots, 2k - 1,$$

and

$$\mathbf{v}_{2k} = (C_0^{2k}, C_1^{2k}, \ldots, C_{2k-1}^{2k}).$$

Then we have for $2 \leq j \leq 2k - 1$,

$$\langle \mathbf{u}, \mathbf{v}_j \rangle = 0$$

and

$$\langle \mathbf{u}, \mathbf{v}_{2k} \rangle = 2k + 1.$$

Here $\langle \mathbf{x}, \mathbf{y} \rangle$ *denotes the inner product of two vectors* $\mathbf{x}$ *and* $\mathbf{y}$ *in* $\mathbb{R}^{2k}$.

Proof. For an integer p with $0 \leq p \leq 2k - 3$, we let

$$\mathbf{w}_p = (\{0\}^p, 1, 2, 1, \{0\}^{2k-p-3}).$$

Obviously, we have

$$\langle \mathbf{u}, \mathbf{w}_p \rangle = 0$$

for all p. In light of Pascal triangle for binomial coefficients, we conclude that $\mathbf{v}_j$, $j = 2, 3, \ldots, 2k - 1$, are linear combinations of $\mathbf{w}_j$, $j = 0, 1, \ldots, 2k - 3$. Indeed, we have

$$\mathbf{v}_3 = \mathbf{w}_0 + \mathbf{w}_1,$$
$$\mathbf{v}_4 = \mathbf{w}_0 + 2\mathbf{w}_1 + \mathbf{w}_2,$$

$$\vdots$$

$$\mathbf{v}_{2k-1} = C_0^{2k-3}\mathbf{w}_0 + C_1^{2k-3}\mathbf{w}_1 + \cdots + C_{2k-3}^{2k-3}\mathbf{w}_{2k-3}.$$

Thus we have for $2 \leq j \leq 2k - 1$,

$$\langle \mathbf{u}, \mathbf{v}_j \rangle = 0.$$

The vector $\mathbf{v}_{2k}$ is not a linear combination of $\mathbf{w}_j$, $j = 0, 1, \ldots, 2k - 3$. However, the difference

$$\mathbf{v}_{2k} - (\{0\}^{2k-2}, 1, 2)$$

is indeed a linear combination of $\mathbf{w}_j$, $j = 0, 1, \ldots, 2k - 3$, and hence

$$\langle \mathbf{u}, \mathbf{v}_{2k} \rangle = \left\langle \mathbf{u}, (\{0\}^{2k-2}, 1, 2) \right\rangle = 2k + 1. \qquad \square$$

To obtain the evaluation of $\zeta(\{1\}^{2k}, 2n + 1)$, we consider the weighted alternating sum

$$(2k - 1)\{G_{2n+1}(2k - 1, 1) + G_{2n+1}(0, 2k)\}$$
$$- (2k - 3)\{G_{2n+1}(2k - 2, 2) + G_{2n+1}(1, 2k - 1)\}$$
$$+ \cdots + (-1)^k \{G_{2n+1}(k, k) + G_{2n+1}(k - 1, k + 1)\}.$$

By Proposition 6.4.3, such sum is equal to

$$\sum_{p=0}^{k-1} (-1)^p (2k - 2p - 1) \sum_{\alpha=2}^{2n} (-1)^\alpha \zeta(\{1\}^p, 2n + 2 - \alpha) \zeta(\{1\}^{2k-p-1}, \alpha).$$

On the other hand, by Proposition 6.4.5, it is equal to

$$(2k + 1)\zeta(\{1\}^{2k}, 2n + 1) + (2k - 1)\zeta(2n + 2k + 1) + S_2.$$

Thus our assertion follows.

Part 4. The evaluation of $\zeta(\{1\}^{2k+1}, 2n)$

To get a formula for $\zeta(\{1\}^{2k+1}, 2n)$, we need another simple fact about binomial coefficients which can be proved with a similar argument as Proposition 6.4.5.

Proposition 6.4.6. *For each positive integer $k \geq 1$, we let*

$$\mathbf{u} = (-k, (k - 1), \ldots, (-1)^{k+1}, 0, (-1)^k, \ldots, -(k - 1), k),$$

$$\mathbf{v}_j = (C_0^j, C_1^j, \ldots, C_j^j, \{0\}^{2k-j}), \quad j = 2, 3, \ldots, 2k,$$

and

$$\mathbf{v}_{2k} = (C_0^{2k+1}, C_1^{2k+1}, \ldots, C_{2k}^{2k+1})$$

be vectors in $\mathbb{R}^{2k+1}$. Then for $2 \leq j \leq 2k$,

$$\langle \mathbf{u}, \mathbf{v}_j \rangle = 0$$

and

$$\langle \mathbf{u}, \mathbf{v}_{2k+1} \rangle = k + 1.$$

Now we are ready to evaluate $\zeta(\{1\}^{2k+1}, 2n)$. The weighted alternating sum

$$\sum_{p=0}^{k-1} (-1)^p (k-p) \{G_{2n}(m-p-1, p+1) - G_{2n}(p, 2k-p+1)\}$$

is equal to

$$\sum_{p=0}^{k-1} (-1)^p (k-p) \sum_{\alpha=2}^{2n-1} (-1)^{\alpha+1} \zeta(\{1\}^p, 2n+1-\alpha) \zeta(\{1\}^{2k-p}, \alpha)$$

by Proposition 6.4.5. On the other hand, it is equal to

$$(k+1)\zeta(\{1\}^{2k+1}, 2n) - \{k\zeta(2n+2k+1) + S_2\}$$

by Proposition 6.4.6 and decompositions of $G'_{2n}(p, q)$ with $p+q = 2k+1$. Therefore, we get our evaluation of $\zeta(\{1\}^{2k+1}, 2n)$.

6.5. Drinfeld Integrals

There are integral representations for multiple zeta values in terms of *Drinfeld integrals* due to Kontsevich [11, 44], namely

$$\zeta(\alpha_1, \alpha_2, \ldots, \alpha_r)$$
$$= \int_{0 < t_1 < t_2 < \cdots < t_{|\alpha|} < 1} \left(\frac{dt_1}{1-t_1} \prod_{j_1=2}^{\alpha_1} \frac{dt_{j_1}}{t_{j_1}} \right) \left(\frac{dt_{\alpha_1+1}}{1-t_{\alpha_1+1}} \prod_{j_2=\alpha_1+2}^{\alpha_1+\alpha_2} \frac{dt_{j_2}}{t_{j_2}} \right)$$
$$\times \cdots \times \left(\frac{dt_{|\alpha|_{r-1}+1}}{1-t_{|\alpha|_{r-1}+1}} \prod_{j_r=|\alpha|_{r-1}+2}^{|\alpha|} \frac{dt_{j_r}}{t_{j_r}} \right),$$

where $|\alpha|_{r-1} = \alpha_1 + \alpha_2 + \cdots + \alpha_{r-1}$. In particular, for positive integers m and n, we have

$$\zeta(\{1\}^{m-1}, n+1) = \int_{0 < t_1 < \cdots < t_{m+n} < 1} \prod_{j=1}^{m} \frac{dt_j}{1-t_j} \prod_{k=m+1}^{m+n} \frac{dt_k}{t_k}.$$

Drinfeld integrals are convenient to express a sum of multiple zeta values and the number of variables can be reduced. For example, we have

$$\zeta(\{1\}^{m-1}, n+1)$$

$$= \frac{1}{(m-1)!(n-1)!} \int_{1<t_1<t_{m+1}<1} \left(\log \frac{1-t_1}{1-t_{m+1}}\right)^{m-1} \left(\log \frac{1}{t_{m+1}}\right)^{n-1}$$

$$\times \frac{dt_1 dt_{m+1}}{(1-t_1)t_{m+1}}$$

by integrating with respect to the variables $t_2, t_3, \ldots, t_m, t_{m+2}, \ldots, t_{m+n}$. The following proposition first given in [29] is powerful to express a sum of multiple zeta values into a double integral.

Proposition 6.5.1. *For a nonnegative integer p and positive integers q, m and n with $m \geq q$, we have*

$$\sum_{|\alpha|=m} \zeta(\{1\}^p, \alpha_1, \alpha_2, \ldots, \alpha_q + n)$$

$$= \frac{1}{p!(q-1)!(m-q)!(n-1)!} \int_{0<t_1<t_2<1} \left(\log \frac{1}{1-t_1}\right)^p \left(\log \frac{t_2}{t_1}\right)^{m-q}$$

$$\times \left(\log \frac{1-t_1}{1-t_2}\right)^{q-1} \left(\log \frac{1}{t_2}\right)^{n-1} \frac{dt_1 dt_2}{(1-t_1)t_2}.$$

$$(6.5.1)$$

Proof. Let $m = q + \ell$ with $\ell \geq 0$. Consider the integral

$$\int_D \prod_{j=1}^{p+q} \frac{dt_j}{1-t_j} \prod_{i=1}^{\ell} \frac{du_i}{u_i} \prod_{k=p+q+1}^{p+q+n} \frac{dt_k}{t_k}, \qquad (6.5.2)$$

where

$$D: \begin{cases} 0 < t_1 < t_2 < \cdots < t_{p+q+n} < 1, \\ 0 < t_{p+1} < u_1 < u_2 < \cdots < u_\ell < t_{p+q+1} < 1. \end{cases}$$

Note that for any permutations on the sets $\{t_1, t_2, \ldots, t_p\}$ or $\{t_{p+2}, \ldots, t_{p+q}\}$ or $\{u_1, u_2, \ldots, u_\ell\}$ or $\{t_{p+q+2}, \ldots, t_{p+q+n}\}$ yield the same value of (6.5.2), which can be transformed into

$$\frac{1}{p!(q-1)!\ell!(n-1)!} \int_{D_1} \prod_{j=1}^{p+q} \frac{dt_j}{1-t_j} \prod_{i=1}^{\ell} \frac{du_i}{u_i} \prod_{k=p+q+1}^{p+q+n} \frac{dt_k}{t_k},$$

where

$$
D_1 : \begin{cases} 0 < t_1, t_2, \ldots, t_p < t_{p+1} < t_{p+2}, t_{p+3}, \ldots, t_{p+q} < t_{p+q+1} \\ < t_{p+q+2}, t_{p+q+3}, \ldots, t_{p+q+n} < 1, \\ 0 < t_{p+1} < u_1, u_2, \ldots, u_\ell < t_{p+q+1}. \end{cases}
$$

Fixing t_{p+1} and t_{p+q+1} and integrating with respect to the variables

$$
t_1, t_2, \ldots, t_p, t_{p+2}, \ldots, t_{p+q}, u_1, u_2, \ldots, u_\ell, t_{p+q+2}, \ldots, t_{p+q+n},
$$

(6.5.2) is equal to the double integral as shown in (6.5.1).

On the other hand, any permutations on the set $\{u_1, u_2, \ldots, u_\ell\}$ yield the same value of (6.5.2) as

$$
\frac{1}{\ell!} \int_{D_2} \prod_{j=1}^{p+q} \frac{dt_j}{1 - t_j} \prod_{r=1}^{\ell} \frac{du_r}{u_r} \prod_{k=p+q+1}^{p+q+n} \frac{dt_k}{t_k},
$$

where

$$
D_2 : \begin{cases} 0 < t_1 < t_2 < \cdots < t_{p+q} < t_{p+q+1} < t_{p+q+2} < \cdots < t_{p+q+n} < 1, \\ t_{p+1} < u_1, u_2, \ldots, u_\ell < t_{p+q+1}. \end{cases}
$$

Now we want to insert $u_1, u_2, \ldots, u_\ell$ into the sequence $0 < t_1 < t_2 < \cdots < t_{p+q+n} < 1$ one by one to produce multiple zeta values of the form $\zeta(\{1\}^p, \alpha_1, \alpha_2, \ldots, \alpha_q + n)$.

First insert u_1 into the sequence according to

$$
t_{p+1} < u_1 \le t_{p+2}, \quad t_{p+2} < u_1 \le t_{p+3}, \quad \ldots, \quad t_{p+q} < u_1 \le t_{p+q+1}
$$

and hence we divide D_2 into a disjoint union of q subdomains. Once u_1 is inserted, we then insert u_2 into the resulting sequence and divide each subdomain into $q + 1$ subdomains. Continuing this process, we insert $u_1, u_2, \ldots, u_\ell$ one by one into the sequence $0 < t_1 < t_2 < \cdots < t_{p+q+n} < 1$ and between t_{p+1} and t_{p+q+1}. This divides D_2 into a disjoint union of $q(q+1) \cdots (q+\ell-1)$ subdomains and, an integration over each subdomain produces a multiple zeta value of the form $\zeta(\{1\}^p, \alpha_1, \alpha_2, \ldots, \alpha_q + n)$ with $|\alpha| = q + \ell$. So (6.5.2) is transformed into

$$
\frac{1}{\ell!} \sum_{|\alpha|=m} c(\alpha) \zeta(\{1\}^p, \alpha_1, \alpha_2, \ldots, \alpha_q + n),
$$

where $c(\alpha)$ is the number of subdomains of D_2 such that the number of $u_1, u_2, \ldots, u_\ell$ inserted into the interval $T_i = [t_{p+i}, t_{p+i+1}]$ is $\alpha_i - 1$ for every $i = 1, 2, \ldots, q$. Consider the different ways of inserting the variables $u_1, u_2, \ldots, u_\ell$ into $T_1 \cup T_2 \cup \cdots \cup T_q$. For T_1, the possible insertions of $\alpha_1 - 1$ variables from ℓ variables is $\ell(\ell - 1) \cdots (\ell - \alpha_1 + 2)$. Then for T_2, the number of possible insertions of $\alpha_2 - 1$ variables from the remaining $\ell - \alpha_1 + 1$ variables is $(\ell - \alpha_1 + 1) \cdots (\ell - \alpha_1 - \alpha_2 + 3)$. Consequently, we find

$$c(\alpha) = \prod_{i=1}^{|\alpha|-q} (\ell - i + 1) = \ell!,$$

and our assertion follows. $\qquad\square$

As a first step, we express $G_n(p, q)$ as a double integral.

Proposition 6.5.2. *For positive integers $n \geq 2$ and nonnegative integers p and q with $p + q \geq 1$, we have*

$$G_n(p, q) = \frac{1}{p!q!(n-2)!} \int_{0 < t_1 < t_2 < 1} \left(\log \frac{1}{1 - t_1} \right)^p \left(\log \frac{1}{1 - t_2} \right)^q$$
$$\times \left(\log \frac{t_2}{t_1} \right)^{n-2} \frac{dt_1 dt_2}{(1 - t_1)t_2}.$$

Proof. As mentioned in Remark 6.4.4, we rewrite $G_n(p, q)$ as

$$\sum_{\mathbf{k} \in \mathbb{N}^{p+1}} \sum_{\mathbf{j} \in \mathbb{N}^q} \frac{1}{s_1 s_2 \cdots s_p s_{p+1}^{n-1} \sigma_1 \sigma_2 \cdots \sigma_q (s_{p+1} + \sigma_q)},$$

where $s_\ell = k_1 + k_2 + \cdots + k_\ell$ and $\sigma_m = j_1 + j_2 + \cdots + j_m$. Therefore, we have

$$G_n(p, q) = \int_D \prod_{j=1}^{p+1} \frac{dt_j}{1 - t_j} \prod_{k=p+2}^{n+p-1} \frac{dt_k}{t_k} \prod_{\ell=1}^{q} \frac{du_\ell}{1 - u_\ell} \frac{dt_{n+p}}{t_{n+p}},$$

where D is a domain in $\mathbb{R}^{p+q+n}$ defined by

$$0 < t_1 < \cdots < t_{n+p} < 1 \quad \text{and} \quad 0 < u_1 < \cdots < u_q < t_{n+p}.$$

Our assertion then follows from fixing t_{p+1} and t_{n+p} as new variables t_1 and t_2, and then integrating with respect to the remaining variables. $\qquad\square$

Now we are ready to decompose $G_n(p, q)$ into a sum of multiple zeta values through its integral representation.

Theorem 6.5.3. *For positive integers $n \geq 2$ and nonnegative integers p and q with $p + q \geq 1$, we have*

$$G_n(p,q) = \sum_{q_1 + q_2 = q} \binom{p + q_1}{p} \sum_{|\alpha| = n + q_2 - 1} \zeta(\{1\}^{p+q_1}, \alpha_1, \alpha_2, \ldots, \alpha_{q_2+1} + 1)$$

$$= \sum_{q_1 + q_2 = q} \binom{p + q_1}{p} \sum_{|\beta| = n + q_2 - 1} \zeta(\beta_1, \beta_2, \ldots, \beta_{n-1} + p + q_1 + 1)$$

$$= \sum_{q_1 = 0}^{q} \binom{p + q_1}{p} \sum_{|c| = p + q + 1} \zeta(c_1, c_2, \ldots, c_{p+q_1+1} + n - 1).$$

Proof. We begin with the integral representation of $G_n(p,q)$, given by

$$G_n(p,q) = \frac{1}{p!q!(n-2)!} \int_{0 < t_1 < t_2 < 1} \left(\log \frac{1}{1 - t_1} \right)^p \left(\log \frac{1}{1 - t_2} \right)^q$$
$$\times \left(\log \frac{t_2}{t_1} \right)^{n-2} \frac{dt_1 dt_2}{(1 - t_1)t_2}.$$

We regard the factor $(-\log(1 - t_2))^q$ as

$$\left(\log \frac{1}{1 - t_1} + \log \frac{1 - t_1}{1 - t_2} \right)^q = \sum_{q_1 + q_2 = q} \frac{q!}{q_1! q_2!} \left(\log \frac{1}{1 - t_1} \right)^{q_1} \left(\log \frac{1 - t_1}{1 - t_2} \right)^{q_2},$$

so

$$G_n(p,q) = \sum_{q_1 + q_2 = q} \frac{1}{q_1! q_2! p!(n-2)!} \int_{0 < t_1 < t_2 < 1} \left(\log \frac{1}{1 - t_1} \right)^{p+q_1} \left(\log \frac{1 - t_1}{1 - t_2} \right)^{q_2}$$
$$\times \left(\log \frac{t_2}{t_1} \right)^{n-2} \frac{dt_1 dt_2}{(1 - t_1)t_2}$$

$$= \sum_{q_1 + q_2 = q} \binom{p + q_1}{p} \sum_{|\alpha| = n + q_2 - 1} \zeta(\{1\}^{p+q_1}, \alpha_1, \alpha_2, \ldots, \alpha_{q_2+1} + 1)$$

by Proposition 6.5.1. Under the change of variables

$$t_1 = 1 - u_2, \quad t_2 = 1 - u_1, \quad 0 < u_1 < u_2 < 1,$$

we get another integral representation for $G_n(p,q)$ as

$$\sum_{q_1 + q_2 = q} \frac{1}{q_1! q_2! p!(n-2)!} \int_{0 < u_1 < u_2 < 1} \left(\log \frac{1 - u_1}{1 - u_2} \right)^{n-2} \left(\log \frac{u_2}{u_1} \right)^{q_2}$$
$$\times \left(\log \frac{1}{u_2} \right)^{p+q_1} \frac{du_1 du_2}{(1 - u_1)u_2}$$

which is equal to

$$\sum_{q_1+q_2=q} \binom{p+q_1}{p} \sum_{|\beta|=n+q_2-1} \zeta(\beta_1, \beta_2, \ldots, \beta_{n-1} + p + q_1 + 1).$$

Finally, the above sum is also equal to

$$\sum_{q_1+q_2=q} \binom{p+q_1}{p} \sum_{|c|=p+q+1} \zeta(c_1, c_2, \ldots, c_{p+q_1+1} + n - 1)$$

by Ohno's duality theorem and sum formula [44]. □

Remark 6.5.4. For a vector

$$\mathbf{k} = \left(\{1\}^{a_1-1}, b_1 + 1, \{1\}^{a_2-1}, b_2 + 1, \ldots, \{1\}^{a_r-1}, b_r + 1 \right)$$

with positive integers $a_1, b_1, a_2, b_2, \ldots, a_r, b_r$, the dual of $\mathbf{k}$ is given by

$$\mathbf{k}' = \left(\{1\}^{b_r-1}, a_r + 1, \ldots, \{1\}^{b_2-1}, a_2 + 1, \{1\}^{b_1-1}, a_1 + 1 \right).$$

Then *Ohno's duality theorem* and sum formula asserted that

$$\sum_{|\alpha|=\ell} \zeta(\mathbf{k} + \alpha) = \sum_{|\beta|=\ell} \zeta(\mathbf{k}' + \beta)$$

for any nonnegative integer ℓ. We employ Ohno's theorem with $\ell = q_2$, $\mathbf{k} = \left(\{1\}^{n-2}, p + q_1 + 2 \right)$ and $\mathbf{k}' = \left(\{1\}^{p+q_1}, n \right)$ in the preceding theorem.

In the following, we proceed to prove the sum formula and restricted sum formula.

For positive integers m, j and n with $n \geq 2$, we let

$$S_j = \sum_{|\alpha|=m+1} \zeta(\alpha_1, \alpha_2, \ldots, \alpha_j + n - 1), \quad j = 1, 2, \ldots, m.$$

By the decomposition of $G_n(p, q)$ with $p + q = m$, we are able to rewrite $G_n(p, q)$ as a linear combination of $F = \zeta(\{1\}^m, n)$ and S_j as follows:

$$
\begin{aligned}
G_n(0, m) &= & F + & S_m + & S_{m-1} + \cdots + S_3 + S_2 + S_1, \\
G_n(1, m - 1) &= & C_1^m F + C_1^{m-1} S_m + C_1^{m-2} S_{m-1} + \cdots + 2S_3 + S_2, \\
G_n(2, m - 2) &= & C_2^m F + C_2^{m-1} S_m + C_2^{m-2} S_{m-1} + \cdots + S_3, \\
& & \cdots \cdots \cdots \cdots \cdots \cdots \cdots \\
G_n(m - 2, 2) &= & C_{m-2}^m F + C_{m-2}^{m-1} S_m + & S_{m-1}, \\
G_n(m - 1, 1) &= & C_{m-1}^m F + & S_m, \\
G_n(m, 0) &= & F,
\end{aligned}
$$

where $C_k^n = \binom{n}{k}$. We see immediately

$$\sum_{p+q=m} (-1)^p G_n(p,q) = \zeta(m+n), \qquad (6.5.3)$$

which is equivalent to the well-known sum formula.

Theorem 6.5.5 (Sum Formula). *For all positive integers m and n with $n \geq 2$, we have*

$$\sum_{|\alpha|=m+n-1} \zeta(\alpha_1, \alpha_2, \ldots, \alpha_{m+1} + 1) = \zeta(m+n). \qquad (6.5.4)$$

Proof. By Proposition 6.5.1, the sum is equal to the double integral

$$\frac{1}{m!(n-2)!} \int_{0<t_1<t_2<1} \left(\log \frac{1-t_1}{1-t_2}\right)^m \left(\log \frac{t_2}{t_1}\right)^{n-2} \frac{dt_1 dt_2}{(1-t_1)t_2}.$$

Replacing the factor $\left(\log \frac{1-t_1}{1-t_2}\right)^m$ by

$$\left(-\log \frac{1}{1-t_1} + \log \frac{1}{1-t_2}\right)^m = \sum_{p+q=m} \frac{m!}{p!q!}(-1)^p \left(\log \frac{1}{1-t_1}\right)^p \left(\log \frac{1}{1-t_2}\right)^q,$$

we obtain

$$\sum_{|\alpha|=m+n-1} \zeta(\alpha_1, \alpha_2, \ldots, \alpha_{m+1} + 1) = \sum_{p+q=m} (-1)^p G_n(p,q).$$

So our assertion follows by (6.5.3). $\qquad \square$

Theorem 6.5.6 (Restricted Sum Formula). *For a nonnegative integer p and positive integers m and q with $m \geq q$, we have*

$$\sum_{|\alpha|=m} \zeta(\{1\}^p, \alpha_1, \alpha_2, \alpha_q + 1) = \sum_{|c|=p+q} \zeta(c_1, c_2, \ldots, c_{p+1} + m - q + 1). \quad (6.5.5)$$

Proof. Let R be the sum of multiple zeta values on the left side of the identity. By Proposition 6.5.1, we have

$$R = \frac{1}{p!(q-1)!(m-q)!} \int_{0<t_1<t_2<1} \left(\log \frac{1}{1-t_1}\right)^p \left(\log \frac{1-t_1}{1-t_2}\right)^{q-1}$$
$$\times \left(\log \frac{t_2}{t_1}\right)^{m-q} \frac{dt_1 dt_2}{(1-t_1)t_2}.$$

Rewriting $\left(\log \frac{1-t_1}{1-t_2}\right)^{q-1}$ as its binomial expansion

$$\sum_{q_1+q_2=q-1} \frac{(q-1)!}{q_1!q_2!} (-1)^{q_1} \left(\log \frac{1}{1-t_1}\right)^{q_1} \left(\log \frac{1}{1-t_2}\right)^{q_2},$$

we get

$$R = \sum_{q_1+q_2=q-1} (-1)^{q_1} \binom{p+q_1}{p} G_{m-q+2}(p+q_1+1, q_2)$$

$$= \sum_{q_1+q_2=q-1} (-1)^{q_1} \binom{p+q_1}{p} \sum_{q_{21}+q_{22}=q_2} \binom{p+q_1+q_{21}}{p+q_1}$$

$$\times \sum_{|\mathbf{c}|=p+q} \zeta(c_1, c_2, \ldots, c_{p+q_1+q_{21}+1} + m - q + 1)$$

by Theorem 6.5.3. Let $k = q_1 + q_{21}$ be a new dummy variable in place of q_{21}. Then we have

$$R = \sum_{k+q_{22}=q-1} \binom{p+k}{p} \sum_{q_1=0}^{k} (-1)^{q_1} \binom{k}{q_1}$$

$$\times \sum_{|\mathbf{c}|=p+q} \zeta(c_1, c_2, \ldots, c_{p+k+1} + m - 1 + 1).$$

Note that the second summation over q_1 will vanish unless $k = 0$ and hence $q_{22} = q - 1$. This completes our assertion.   □

6.6. Exercises

1. By using the identity

$$\zeta\left(ps; x + \frac{1}{2}\right) \zeta(qs; x)$$

$$= \sum_{n_1=0}^{\infty} \sum_{n_2=0}^{\infty} \left[\left(n_1 + n_2 + x + \frac{1}{2}\right)^p (n_2 + x)^q\right]^{-s}$$

$$+ \sum_{n_1=0}^{\infty} \sum_{n_2=0}^{\infty} \left[\left(n_1 + x + \frac{1}{2}\right)^p (n_1 + n_2 + 1 + x)^q\right]^{-s},$$

prove that for all positive integers m and n with $w = 2m + 2n + 1$,

$$B_{2m}(x)B_{2n+1}\left(x + \frac{1}{2}\right)$$

$$= 2m \sum_{k=0}^{n} \binom{2n+1}{2k} B_{2k}\left(\frac{1}{2}\right) \frac{B_{w-2k}(x)}{w - 2k}$$

$$+ (2n+1) \sum_{k=0}^{m} \binom{2m}{2k} B_{2k}\left(\frac{1}{2}\right) \frac{B_{w-2k}(x + 1/2)}{w - 2k}$$

and

$$S_{2m,2n+1}^{+-} = \frac{1}{2}\eta(w) + \sum_{r=0}^{n} \binom{w - 2r - 1}{2m - 1} \eta(2k)\zeta(w - 2k)$$

$$- \sum_{r=0}^{m} \binom{w - 2r - 1}{2n} \eta(2k)\eta(w - 2k).$$

2. Prove that for positive integers m, k and j with $1 \le j < k$,

$$\sum_{r=0}^{m} \binom{m}{r} B_r\left(\frac{j}{k}\right)\left(x - \frac{j}{k}\right)^{m-r} = B_m(x).$$

3. For all positive integers m, n and k with $w = 2m + 2n + 1$, prove that

$$T_{2m,2n+1}^{(k)} = \frac{1}{2}k^{-(2n+1)}\zeta(w) + \sum_{r=0}^{m} \binom{w - 2r - 1}{2n} k^{2m-2r}\zeta(2r)\zeta(w - 2r)$$

$$+ k^{2m-1} \sum_{r=0}^{n} \binom{w - 2r - 1}{2m - 1} \zeta(2r)\zeta(w - 2r)$$

$$+ k^{2m-1} \sum_{j=1}^{k-1}\sum_{r=0}^{n} \binom{w - 2r - 1}{2m - 1} C_{2r}\left(\frac{j}{k}\right) C_{w-2r}\left(\frac{j}{k}\right)$$

$$+ k^{2m-1} \sum_{j=1}^{k-1}\sum_{r=0}^{n} \binom{w - 2r - 2}{2m - 1} S_{2r+1}\left(\frac{j}{k}\right) S_{w-2r-1}\left(\frac{j}{k}\right).$$

4. If we represent a multiple zeta value by the form

$$\int_0^1 \Omega_1\Omega_2 \cdots \Omega_n,$$

where $\Omega_j = dt_j/(1-t_j)$ or dt_j/t_j and $\Omega_1 = dt_1/(1-t_1)$, $\Omega_n = dt_n/t_n$, the *weight-length shuffles* take the form

$$\int_0^1 \Omega_1\Omega_2\cdots\Omega_n \int_0^1 \Omega_{n+1}\Omega_{n+2}\cdots\Omega_{n+m}$$
$$= \sum_\sigma \int_0^1 \Omega_{\sigma(1)}\Omega_{\sigma(2)}\cdots\Omega_{\sigma(n+m)},$$

where the sum is over all $\binom{n+m}{n}$ permutations σ of the set $\{1, 2, \ldots, n+m\}$ which preserve the relative orders of 1-forms $\Omega_1, \Omega_2, \ldots, \Omega_{n+m}$. More precisely, for all $1 \le i < j \le n$ and $n+1 \le i < j \le n+m$, we have

$$\sigma^{-1}(i) < \sigma^{-1}(j).$$

Employ the weight-length shuffles to prove that

$$\zeta(2)\zeta(1, 2) = 6\zeta(1, 1, 3) + 3\zeta(1, 2, 2) + \zeta(2, 1, 2).$$

5. The *depth-length shuffle formula* produced from the product of two multiple zeta values

$$\zeta(\alpha_1, \alpha_2, \ldots, \alpha_r) = \sum_{1 \le n_1 < n_2 < \cdots < n_r} n_1^{-\alpha_1} n_2^{-\alpha_2} \cdots n_r^{-\alpha_r}$$

and

$$\zeta(\beta_1, \beta_2, \ldots, \beta_\ell) = \sum_{1 \le m_1 < m_2 < \cdots < m_\ell} m_1^{-\beta_1} m_2^{-\beta_2} \cdots m_r^{-\beta_\ell}$$

is the sum over possible insertions of $m_1, m_2, \ldots, m_\ell$ into the sequence $1 \le n_1 < n_2 < \cdots < n_r$ which preserve the orders of $m_1, m_2, \ldots, m_\ell$. With the depth-length shuffle, prove that

$$\zeta(1, 2)\zeta(2) = 2\zeta(1, 2, 2) + \zeta(2, 1, 2) + \zeta(3, 2) + \zeta(1, 4).$$

Part II

Theory of Modular Forms of Several Variables

Chapter 7

Theory of Modular Forms of Several Variables

The theory of modular forms of several variables was initiated by Carl Ludwig Siegel around 1945 as a generalization of the classical modular forms of one variable. Based on the book "Siegelsche Modulfunktionen" by E. Freitag, we describe the generalized upper half-plane, the real symplectic group and the full modular group in details and then the theory.

7.1. The Generalized Upper Half-Plane

Let $M_n(\mathbb{R})$ be the matrix ring over real numbers of size n. The *generalized upper half-plane* or so-called *Siegel upper half-plane* is defined by

$$\mathcal{H}_n = \left\{ Z = X + iY \mid {}^tX = X \in M_n(\mathbb{R}), {}^tY = Y \in M_n(\mathbb{R}), Y > 0 \right\},$$

where ${}^tX$ denotes the transpose of X. Note that $\mathcal{H}_n$ is a typical tube domain in $\mathbb{C}^{n(n+1)/2}$. A point Z in $\mathcal{H}_n$ is an $n \times n$ symmetric matrix over $\mathbb{C}$ with a positive definite imaginary part. Such a matrix is always invertible as shown below.

Proposition 7.1.1. *For each Z in $\mathcal{H}_n$, Z is invertible as an $n \times n$ matrix over $\mathbb{C}$ and furthermore, $-Z^{-1} \in \mathcal{H}_n$.*

Proof. Suppose that Z is singular. Then as a linear transform from $\mathbb{C}^n$ to $\mathbb{C}^n$, the transformation is not one-to-one. So there exists an $\mathbf{u} \in \mathbb{C}^n$, $\mathbf{u} \neq \mathbf{0}$, such that

$$Z\mathbf{u} = \mathbf{0}.$$

It follows that ${}^t\overline{\mathbf{u}}Z\mathbf{u} = 0$. Set $Z = X + iY$ as in the definition of $\mathcal{H}_n$. Then we have

$$ {}^t\overline{\mathbf{u}}X\mathbf{u} + i\,{}^t\overline{\mathbf{u}}Y\mathbf{u} = 0.$$

Both ${}^t\overline{\mathbf{u}}X\mathbf{u}$ and ${}^t\overline{\mathbf{u}}Y\mathbf{u}$ are real numbers, it forces that

$$ {}^t\overline{\mathbf{u}}X\mathbf{u} = 0 \quad \text{and} \quad {}^t\overline{\mathbf{u}}Y\mathbf{u} = 0.$$

A contradiction to the fact that Y is positive definite. Note that the imaginary part of $-Z^{-1}$ is given by

$$
\begin{aligned}
\frac{1}{2i}\left(-Z^{-1} - \left(-\overline{Z}^{-1}\right)\right) &= \frac{1}{2i}\left(\overline{Z}^{-1} - Z^{-1}\right) \\
&= \frac{1}{2i}\left(\overline{Z}^{-1}\left(Z - \overline{Z}\right)Z^{-1}\right) \\
&= Y\left[Z^{-1}\right].
\end{aligned}
$$

As Z^{-1} is nonsingular, $Y[Z^{-1}]$ is positive definite provided that Y is positive definite. This proves that $-Z^{-1} \in \mathcal{H}_n$. $\square$

Remark 7.1.2. Here we prove that $-Z^{-1}$ is symmetric. It follows directly from

$$ {}^t(-Z^{-1}) = -{}^tZ^{-1} = -Z^{-1}.$$

Remark 7.1.3. A symmetric $n \times n$ matrix S over $\mathbb{R}$ is positive definite if $S[\mathbf{u}] = {}^t\overline{\mathbf{u}}S\mathbf{u} > 0$ for all $\mathbf{u} \in \mathbb{C}^n$, $\mathbf{u} \neq \mathbf{0}$. This is equivalent to say that all eigenvalues of S are positive or there exists a positive number λ such that

$$ S[\mathbf{u}] \geq \lambda|\mathbf{u}|^2$$

for all $\mathbf{u} \in \mathbb{C}^n$.

For a positive definite matrix S, it has a square root $S^{1/2}$ which is also symmetric and satisfies

$$ (S^{1/2})(S^{1/2}) = S.$$

Indeed, if $S = D[U]$ with $D = \text{diag}\,[\lambda_1, \lambda_2, \ldots, \lambda_n]$ and U being orthonormal, i.e., ${}^tUU = E$, then $S^{1/2} = D^{1/2}[U]$ with

$$D^{1/2} = \text{diag}\left[\sqrt{\lambda_1}, \sqrt{\lambda_2}, \ldots, \sqrt{\lambda_n}\right].$$

As a typical tube domain in $\mathbb{C}^{n(n+1)/2}$, $\text{Aut}(\mathcal{H}_n)$, the set of bi-holomorphic mappings from $\mathcal{H}_n$ into itself is generated by the following mappings.

(1) $p_B\colon Z \mapsto Z + B$, ${}^tB = B \in M_n(\mathbb{R})$. The mapping p_B is called a translation.

(2) $t_U\colon Z \mapsto Z[U] = {}^tUZU$, $U \in \text{GL}_n(\mathbb{R})$. The mapping t_U is called a dilation.

(3) $\iota\colon Z \mapsto -Z^{-1}$. Such a mapping is called the involution of $\mathcal{H}_n$.

Remark 7.1.4. All the mappings mention above have their representatives as $2n \times 2n$ matrices over $\mathbb{R}$. The matrix representation of p_B is

$$\begin{bmatrix} E & B \\ 0 & E \end{bmatrix}$$

while the matrix representation of t_U is

$$\begin{bmatrix} {}^tU & 0 \\ 0 & U^{-1} \end{bmatrix}.$$

Also the matrix representation of the involution ι is given by

$$\begin{bmatrix} 0 & -E \\ E & 0 \end{bmatrix}.$$

These matrices are special examples of real symplectic matrices which act on the generalized upper half-plane $\mathcal{H}_n$ by the action

$$M = \begin{bmatrix} A & B \\ C & D \end{bmatrix} \colon Z \mapsto (AZ + B)(CZ + D)^{-1}.$$

7.2. The Real Symplectic Group

The *real symplectic group of degree n* is defined as

$$\text{Sp}(n, \mathbb{R}) = \left\{ M \in M_{2n}(\mathbb{R}) \,\middle|\, {}^tMJM = J, J = \begin{bmatrix} 0 & -E \\ E & 0 \end{bmatrix} \right\}.$$

If we decompose a $2n \times 2n$ matrix M into four $n \times n$ submatrices as

$$M = \begin{bmatrix} A & B \\ C & D \end{bmatrix},$$

then we have

$$M \in \mathrm{Sp}(n, \mathbb{R}) \quad \text{if and only if} \quad M^{-1} = J^{-1t}MJ = \begin{bmatrix} {}^t D & -{}^t B \\ -{}^t C & {}^t A \end{bmatrix}.$$

Proposition 7.2.1. *The following statements hold.*

(1) *A matrix* $M = \begin{bmatrix} A & B \\ C & D \end{bmatrix}$ *in* $M_{2n}(\mathbb{R})$ *is symplectic if and only if*

$${}^t AD - {}^t CB = E, \quad {}^t AC = {}^t CA, \quad {}^t BD = {}^t DB.$$

(2) *If* M *is symplectic, then* ${}^t M$ *is also symplectic and hence*

$$A^t D - B^t C = E, \quad A^t B = B^t A, \quad C^t D = D^t C.$$

(3) *The inverse of a symplectic matrix* M *is given by*

$$M^{-1} = J^{-1t}MJ = \begin{bmatrix} {}^t D & -{}^t B \\ -{}^t C & {}^t A \end{bmatrix}.$$

(4) *Examples of symplectic matrices are as follows:*

(a) $\begin{bmatrix} E & S \\ 0 & E \end{bmatrix}, \ {}^t S = S \in M_n(\mathbb{R})$;

(b) $\begin{bmatrix} {}^t U & 0 \\ 0 & U^{-1} \end{bmatrix}, \ U \in \mathrm{GL}_n(\mathbb{R})$;

(c) $J = \begin{bmatrix} 0 & -E \\ E & 0 \end{bmatrix}$.

Proposition 7.2.2. *The real symplectic group* $\mathrm{Sp}(n, \mathbb{R})$ *is generated by the following elements.*

(a) $\begin{bmatrix} E & S \\ 0 & E \end{bmatrix}, \ {}^t S = S \in M_n(\mathbb{R})$;

(b) $\begin{bmatrix} {}^t U & 0 \\ 0 & U^{-1} \end{bmatrix}, \ U \in \mathrm{GL}_n(\mathbb{R})$;

(c) $J = \begin{bmatrix} 0 & -E \\ E & 0 \end{bmatrix}$.

In particular, if $M \in \mathrm{Sp}(n, \mathbb{R})$, *then* $\det M = 1$.

Proof. Suppose that $M = \begin{bmatrix} A & B \\ C & D \end{bmatrix}$ is a symplectic matrix. If $C = 0$, then $A^t D = E$ and

$$M = \begin{bmatrix} A & 0 \\ 0 & {}^tA^{-1} \end{bmatrix} \begin{bmatrix} E & A^{-1}B \\ 0 & E \end{bmatrix}.$$

From $A^t B = B^t A$, we conclude that $A^{-1}B$ is a symmetric matrix. So M is indeed a product of an element of (b) and an element of (a).

On the other hand, if $\det C \neq 0$, then

$$\begin{bmatrix} A & B \\ C & D \end{bmatrix} \begin{bmatrix} E & -C^{-1}D \\ 0 & E \end{bmatrix} \begin{bmatrix} 0 & -E \\ E & 0 \end{bmatrix} = \begin{bmatrix} * & * \\ 0 & * \end{bmatrix}.$$

By the previous discussion, we also have that M is a product of elements in (a), (b) and (c).

Finally, we suppose that $1 \leq \operatorname{rank} C = r < n$, so there exist two matrices U and V in $\operatorname{GL}_n(\mathbb{R})$ such that

$$UCV = \begin{bmatrix} C_r & 0 \\ 0 & 0 \end{bmatrix} \quad \text{with} \quad \det C_r \neq 0.$$

It follows that

$$\begin{bmatrix} {}^tU^{-1} & 0 \\ 0 & U \end{bmatrix} \begin{bmatrix} A & B \\ C & D \end{bmatrix} \begin{bmatrix} V & 0 \\ 0 & {}^tV^{-1} \end{bmatrix} = \begin{bmatrix} * & * \\ UCV & * \end{bmatrix}.$$

The resulting matrix appears to be the form

$$\begin{bmatrix} A_r & 0 & B_r & * \\ * & {}^tU^{-1} & * & * \\ C_r & 0 & D_r & * \\ 0 & 0 & 0 & U \end{bmatrix} = \begin{bmatrix} A_r & 0 & B_r & 0 \\ 0 & E_{n-r} & 0 & 0 \\ C_r & 0 & D_r & 0 \\ 0 & 0 & 0 & E_{n-r} \end{bmatrix} \begin{bmatrix} * & * \\ 0 & * \end{bmatrix},$$

where

$$\begin{bmatrix} A_r & B_r \\ C_r & D_r \end{bmatrix} \in \operatorname{Sp}(r, \mathbb{R}).$$

Note that

$$\begin{bmatrix} A_r & 0 & B_r & 0 \\ 0 & E_{n-r} & 0 & 0 \\ C_r & 0 & D_r & 0 \\ 0 & 0 & 0 & E_{n-r} \end{bmatrix} \begin{bmatrix} E & 0 \\ \epsilon E & E \end{bmatrix} = \begin{bmatrix} A_r + \epsilon B_r & 0 & B_r & 0 \\ 0 & E_{n-r} & 0 & 0 \\ C_r + \epsilon D_r & 0 & D_r & 0 \\ 0 & \epsilon E_{n-r} & 0 & E_{n-r} \end{bmatrix}.$$

There exists a sufficiently small $\epsilon > 0$ so that $\det(C_r + \epsilon D_r) \neq 0$. So our assertion then follows from the cases when $\det C \neq 0$. $\square$

Remark 7.2.3. For each positive integer r with $r < n$, $\mathrm{Sp}(r, \mathbb{R})$ can be viewed as a subgroup of $\mathrm{Sp}(n, \mathbb{R})$ via the embedding

$$
M_r = \begin{bmatrix} A_r & B_r \\ C_r & D_r \end{bmatrix} \hookrightarrow M = \begin{bmatrix} A_r & 0 & B_r & 0 \\ 0 & E_{n-r} & 0 & 0 \\ C_r & 0 & D_r & 0 \\ 0 & 0 & 0 & E_{n-r} \end{bmatrix}.
$$

7.3. The Action of the Symplectic Group

The real symplectic group $\mathrm{Sp}(n, \mathbb{R})$ acts on $\mathcal{H}_n$ by the action

$$
M = \begin{bmatrix} A & B \\ C & D \end{bmatrix} : Z \mapsto M(Z) = (AZ + B)(CZ + D)^{-1}.
$$

In particular, the matrices corresponding to the translation $p_B \colon Z \mapsto Z + B$ and the dilation $t_U \colon Z \mapsto Z[U]$ are $\left[\begin{smallmatrix} E & B \\ 0 & E \end{smallmatrix}\right]$ and $\left[\begin{smallmatrix} {}^t U & 0 \\ 0 & U^{-1} \end{smallmatrix}\right]$, respectively. Also the matrix corresponding to the involution $\iota \colon Z \mapsto -Z^{-1}$ is

$$
J = \begin{bmatrix} 0 & -E \\ E & 0 \end{bmatrix}.
$$

In the following we shall prove that the action is well-defined.

Proposition 7.3.1. *For a symplectic matrix* $M = \left[\begin{smallmatrix} A & B \\ C & D \end{smallmatrix}\right]$ *and* $Z \in \mathcal{H}_n$, *then* $\det(CZ + D) \neq 0$ *and* $M(Z) = (AZ + B)(CZ + D)^{-1} \in \mathcal{H}_n$.

Proof. First we prove that $\det(CZ + D) \neq 0$. If $C = 0$, then $\det(CZ + D) = \det D \neq 0$ since D must be nonsingular when $C = 0$. Also if $\det C \neq 0$, then $C^{-1}D$ is symmetric since $C^t D = D^t C$. So that $Z + C^{-1}D \in \mathcal{H}_n$ and hence

$$
\det(CZ + D) = \det C \det(Z + C^{-1}D) \neq 0.
$$

For general cases, suppose that $\mathrm{rank}\, C = r$ with $1 \leq r < n$. Then there exist $U, V \in \mathrm{GL}_n(\mathbb{R})$ such that

$$
UCV = \begin{bmatrix} C_r & 0 \\ 0 & 0 \end{bmatrix} \quad \text{and} \quad UD^t V^{-1} = \begin{bmatrix} D_r & * \\ 0 & D_{n-r} \end{bmatrix}
$$

with $\det C_r \neq 0$. Now we have

$$
\begin{aligned}
\det(CZ + D) &= \det \left[U(CZ + D)^t V^{-1} \right] (\det U)^{-1} (\det V) \\
&= \det \left[(UCV)Z[^t V^{-1}] + UD^t V^{-1} \right] (\det U)^{-1} (\det V) \\
&= \det(C_r Z_r + D_r) \det D_{n-r} (\det U)^{-1} (\det V),
\end{aligned}
$$

where

$$
Z[^t V^{-1}] = \begin{bmatrix} Z_r & * \\ * & Z_{n-r} \end{bmatrix}.
$$

Our assertion then follows from an induction on the degree n.

Next we prove that $M(Z)$ is symmetric. It follows from

$$
\begin{aligned}
{}^t M(Z) &- M(Z) \\
&= ({}^t Z^t C + {}^t D)^{-1} ({}^t Z^t A + {}^t B) - (AZ + B)(CZ + D)^{-1} \\
&= ({}^t Z^t C + {}^t D)^{-1} \left[({}^t Z^t A + {}^t B)(CZ + D) - ({}^t Z^t C + {}^t D)(AZ + B) \right] \\
&\quad \times (CZ + D)^{-1} \\
&= 0
\end{aligned}
$$

since ${}^t AC = {}^t CA$, ${}^t BC - {}^t DA = -E$, ${}^t BD = {}^t DB$ and ${}^t AD - {}^t CB = E$.

Finally, we prove that $M(Z)$ has a positive definite imaginary part. Note that

$$
\begin{aligned}
\frac{1}{2i} &\left[{}^t M(Z) - \overline{M(Z)} \right] \\
&= \frac{1}{2i} (Z^t C + {}^t D)^{-1} \left[(Z^t A + {}^t B)(C\overline{Z} + D) - (Z^t C + {}^t D)(A\overline{Z} + B) \right] \\
&\quad \times (C\overline{Z} + D)^{-1} \\
&= \frac{1}{2i} (Z^t C + {}^t D)^{-1} (Z - \overline{Z})(C\overline{Z} + D)^{-1} \\
&= Y \left[(C\overline{Z} + D)^{-1} \right].
\end{aligned}
$$

As Y is positive definite and $(C\overline{Z} + D)^{-1}$ is nonsingular, we conclude that $Y \left[(C\overline{Z} + D)^{-1} \right]$ is also positively definite. Consequently, we have that $M(Z) \in \mathcal{H}_n$ if M is symplectic and $Z \in \mathcal{H}_n$. $\qquad \square$

Let $Z = X + iY \in \mathcal{H}_n$ with

$$
X = [x_{jk}] \quad \text{and} \quad Y = [y_{jk}].
$$

Define the usual Euclidean measures

$$dX = \prod_{j \leq k} dx_{jk} \quad \text{and} \quad dY = \prod_{j \leq k} dy_{jk}.$$

Proposition 7.3.2. *Notation as above. The measure*

$$(\det Y)^{-(n+1)} dX dY$$

is an invariant measure on $\mathcal{H}_n$ *under the action of the real symplectic group* $\mathrm{Sp}(n, \mathbb{R})$.

Proof. Let

$$M = \begin{bmatrix} A & B \\ C & D \end{bmatrix} \in \mathrm{Sp}(n, \mathbb{R}) \quad \text{and} \quad Z_1, Z_2 \in \mathcal{H}_n.$$

Then

$$
\begin{aligned}
M(Z_1) - M(Z_2) &= (Z_1{}^t C + {}^t D)^{-1} (Z_1{}^t A + {}^t B) - (AZ_2 + B)(CZ_2 + D)^{-1} \\
&= (Z_1{}^t C + {}^t D)^{-1} (Z_1 - Z_2)(CZ_2 + D)^{-1}.
\end{aligned}
$$

Set $Z_1 = Z + \Delta Z$ and $Z_2 = Z$. It follows that

$$M(Z + \Delta Z) - M(Z) = \left[(Z + \Delta Z)^t C + {}^t D \right]^{-1} \Delta Z (CZ + D)^{-1}.$$

On the other hand, we also have

$$\operatorname{Im} M(Z) = Y \left[(C\overline{Z} + D)^{-1} \right].$$

Consequently, we have

$$\det(\operatorname{Im} M(Z)) = \det Y |\det(CZ + D)|^{-2}.$$

Also note that under the mapping

$$Z \mapsto Z[\Lambda]$$

with $\Lambda = \operatorname{diag}[\lambda_1, \lambda_2, \ldots, \lambda_n]$. The Jacobian determinant of the transform is $(\det \Lambda)^{2n+2}$. Thus it follows that

$$dM(Z) = |\det(CZ + D)|^{-2(n+1)} dZ$$

and hence

$$(\det \operatorname{Im} M(Z))^{-(n+1)} dM(Z) = (\det Y)^{-(n+1)} dZ.$$

This proves our assertion.   □

Remark 7.3.3. The Jacobian determinant of the transformation $Z \mapsto M(Z)$ is given by $|\det(CZ + D)|^{-2(n+1)}$. It is true for the cases $Z \mapsto Z[\Lambda]$ as we had already proved. Hence it is also true for the mapping $Z \mapsto Z[\Lambda[U]]$ with U being a unitary matrix of size n since the Jacobian determinant of the transformation $Z \mapsto Z[\Lambda]$ is 1. However, the set of matrices of the form $\Lambda[U]$ is dense in the set of nonsingular matrices of size n. Thus we conclude that the Jacobian determinant of the transformation $Z \mapsto M(Z)$ is $|\det(CZ + D)|^{-2(n+1)}$ simply from

$$\Delta M(Z) \approx \Delta Z \left[(C\overline{Z} + D)^{-1} \right].$$

The isotropy subgroup of $\mathrm{Sp}(n, \mathbb{R})$ at $Z = iE$ is given by

$$K = \left\{ \begin{bmatrix} A & B \\ -B & A \end{bmatrix} \,\middle|\, {}^t\!AA + {}^t\!BB = E, {}^t\!AB = {}^t\!BA \right\}$$

and K is isomorphic to $U(n)$, the unitary group of size n, via the mapping

$$\begin{bmatrix} A & B \\ -B & A \end{bmatrix} \mapsto A + iB.$$

Proposition 7.3.4. *The real symplectic group* $\mathrm{Sp}(n, \mathbb{R})$ *acts on* $\mathcal{H}_n$ *transitively by the action*

$$M = \begin{bmatrix} A & B \\ C & D \end{bmatrix} : Z \mapsto M(Z) = (AZ + B)(CZ + D)^{-1}.$$

Furthermore, each symplectic matrix M has the decomposition

$$M = \begin{bmatrix} E & X \\ 0 & E \end{bmatrix} \begin{bmatrix} Y^{1/2} & 0 \\ 0 & Y^{-1/2} \end{bmatrix} \begin{bmatrix} A & B \\ -B & A \end{bmatrix},$$

where ${}^t\!X = X$, Y is a positive definite matrix and $A + iB \in U(n)$.

Proof. Suppose that

$$M(iE) = X + iY$$

and

$$M' = \begin{bmatrix} E & X \\ 0 & E \end{bmatrix} \begin{bmatrix} Y^{1/2} & 0 \\ 0 & Y^{-1/2} \end{bmatrix}.$$

Then $M' \in \mathrm{Sp}(n, \mathbb{R})$ and

$$M(iE) = X + iY = M'(iE).$$

It follows $M'^{-1}M(iE) = iE$ and hence $M'^{-1}M \in K$, the isotropy subgroup of $\mathrm{Sp}(n, \mathbb{R})$ at $Z = iE$. Thus

$$M'^{-1}M = \begin{bmatrix} A & B \\ -B & A \end{bmatrix},$$

and M has the decomposition as asserted. Also for each $Z = X + iY \in \mathcal{H}_n$, we have

$$M'(iE) = Z.$$

Hence the action is transitively. $\square$

7.4. The Generalized Disc

The *generalized disc* D_n is defined by

$$D_n = \left\{ W \in M_n(\mathbb{C}) \mid {}^t W = W, E - {}^t\overline{W}W > 0 \right\}.$$

Observe that D_n is a bounded domain of $\mathbb{C}^{n(n+1)/2}$, which is the image of $\mathcal{H}_n$ under the Cayley transform

$$Z \mapsto (Z - iE)(Z + iE)^{-1}.$$

Proposition 7.4.1. *The Cayley transform maps $\mathcal{H}_n$ onto D_n and its inverse transform is given by*

$$W \mapsto i(E + W)(E - W)^{-1}.$$

Proof. Let

$$W = (Z - iE)(Z + iE)^{-1}.$$

It is a direct verification to prove that ${}^tW = W$. Now we prove $W \in D_n$. A direct calculation shows that

$$
\begin{aligned}
E - {}^t\overline{W}W &= E - (\overline{Z} - iE)^{-1}(\overline{Z} + iE)(Z - iE)(Z + iE)^{-1} \\
&= (\overline{Z} - iE)^{-1}[(\overline{Z} - iE)(Z + iE) - (\overline{Z} + iE)(Z - iE)](Z + iE)^{-1} \\
&= (\overline{Z} - iE)^{-1}[-2i(Z - \overline{Z})](Z + iE)^{-1} \\
&= 4Y\left[(Z + iE)^{-1}\right].
\end{aligned}
$$

Since $Z + iE$ is a nonsingular matrix, it implies that

$$E - {}^t\overline{W}W > 0.$$

For $W \in D_n$, we shall prove that $E - W$ is nonsingular. Suppose that $E - W$ is singular. Then there exists a nonzero vector $\mathbf{u} \in \mathbb{C}^n$ such that

$$(E - W)\mathbf{u} = \mathbf{0}.$$

Then

$$\,^t\overline{\mathbf{u}}(E - {}^t\overline{W}W)\mathbf{u} = {}^t\overline{\mathbf{u}}\mathbf{u} - ({}^t\overline{\mathbf{u}}\,{}^t\overline{W})(W\mathbf{u}) = {}^t\overline{\mathbf{u}}\mathbf{u} - {}^t\overline{\mathbf{u}}\mathbf{u} = 0.$$

A contradiction to the fact that $E - {}^t\overline{W}W > 0$. $\qquad\square$

The matrix corresponding to the Cayley transform is given by

$$M_0 = \frac{1}{\sqrt{2i}} \begin{bmatrix} E & -iE \\ E & iE \end{bmatrix}$$

and its inverse is

$$M_0^{-1} = \frac{1}{\sqrt{2i}} \begin{bmatrix} iE & iE \\ -E & E \end{bmatrix}.$$

For each $M \in \mathrm{Sp}(n, \mathbb{R})$, $M_0 M M_0^{-1}$ has an action on D_n like the action of symplectic matrices on $\mathcal{H}_n$. Indeed, one has

$$M_0 \, \mathrm{Sp}(n, \mathbb{R}) M_0^{-1} = \left\{ \begin{bmatrix} P & Q \\ -\overline{Q} & -\overline{P} \end{bmatrix} \,\Big|\, {}^t P\overline{P} - {}^t\overline{Q}Q = E, {}^t P\overline{Q} = {}^t\overline{Q}P \right\}$$

and the action is given by

$$\begin{bmatrix} P & Q \\ \overline{Q} & \overline{P} \end{bmatrix} : W \mapsto (PW + Q)(\overline{Q}W + \overline{P})^{-1}.$$

In particular, $M_0 \, \mathrm{Sp}(n, \mathbb{R}) M_0^{-1}$ contains the subgroup

$$\left\{ \begin{bmatrix} P & 0 \\ 0 & \overline{P} \end{bmatrix} \,\Big|\, {}^t P\overline{P} = E \right\}$$

which is isomorphic to $U(n)$, the unitary group of size n.

Remark 7.4.2. Besides the generalized disc D_n described as above, there are several bounded domains of interesting in the literature. For example, we have the following Siegel domain of various types.

Type I. $W \in M_n(\mathbb{C})$, $E_n - {}^t\overline{W}W > 0$.

Type II. ${}^tW = -W \in M_n(\mathbb{C})$, $E_n - {}^t\overline{W}W > 0$, $n \geq 3$.

Type III. $W \in M_{p \times q}(\mathbb{C})$, $E_q - {}^t\overline{W}W > 0$.

Especially, the bounded symmetric domain of type I is equivalent to the Hermitian upper half-plane $\mathfrak{H}_n$ defined by

$$\mathfrak{H}_n = \left\{ Z = X + iY \mid {}^t\overline{X} = X \in M_n(\mathbb{C}), {}^t\overline{Y} = Y \in M_n(\mathbb{C}), Y > 0 \right\}$$

which is also a typical tube domain. Here we introduce other tube domain of interest. Let $\mathbb{H}$ be the skew field of quaternions over real numbers, i.e.,

$$\mathbb{H} = \left\{ a + bi + cj + dk \mid a, b, c, d \in \mathbb{R} \right\}.$$

Let

$$\Omega_n = \left\{ Z = X + iY \mid {}^t\overline{X} = X \in M_n(\mathbb{H}), {}^t\overline{Y} = Y \in M_n(\mathbb{H}), Y > 0 \right\}.$$

Then Ω_n is another tube domain contained in $\mathbb{C}^{2n^2+n}$.

7.5. Exercises

1. Find the square root of the positive definite matrix

$$S = \begin{bmatrix} 45 & 44 & 22 \\ 44 & 45 & 22 \\ 22 & 22 & 12 \end{bmatrix}.$$

   Ans. $\begin{bmatrix} 5 & 4 & 2 \\ 4 & 5 & 2 \\ 2 & 2 & 2 \end{bmatrix}$.

2. Suppose that

$$M = \begin{bmatrix} A & B \\ C & D \end{bmatrix} \in \mathrm{Sp}(n, \mathbb{R}) \quad \text{and} \quad \widetilde{M} = \begin{bmatrix} \widetilde{A} & \widetilde{B} \\ \widetilde{C} & \widetilde{D} \end{bmatrix} = \begin{bmatrix} {}^tD & {}^tB \\ {}^tC & {}^tA \end{bmatrix}.$$

   Prove that for $Z_1, Z_2 \in \mathcal{H}_n$,

$$\det\left(M(Z_1) + Z_2 \right) \det\left(CZ_1 + D \right) = \det\left(\widetilde{M}(Z_2) + Z_1 \right) \det\left(\widetilde{C}Z_2 + \widetilde{D} \right).$$

3. The complex symplectic group of degree n is defined as

$$\mathrm{Sp}(n, \mathbb{C}) = \left\{ M \in M_{2n}(\mathbb{C}) \;\middle|\; {}^t\overline{M}JM = J, J = \begin{bmatrix} 0 & -E \\ E & 0 \end{bmatrix} \right\}.$$

Give some typical elements of $\mathrm{Sp}(n, \mathbb{C})$ and a set of generators in $\mathrm{Sp}(n, \mathbb{C})$.

4. Show that the bounded symmetric domain of type I is equivalent to the Hermitian upper half-plane $\mathfrak{H}_n$ defined by

$$\mathfrak{H}_n = \left\{ Z = X + iY \mid {}^t\overline{X} = X \in M_n(\mathbb{C}), {}^t\overline{Y} = Y \in M_n(\mathbb{C}), Y > 0 \right\}$$

under the mapping $Z \mapsto (Z - iE)(Z + iE)^{-1}$.

Chapter 8

The Full Modular Group

The full modular group $\mathrm{Sp}(n, \mathbb{Z})$ is a discrete subgroup of the real symplectic group $\mathrm{Sp}(n, \mathbb{R})$. It is generated by three kinds of symplectic matrices which correspond to translations, dilations and an involution in the generalized upper half-plane.

8.1. The Full Modular Group and Its Subgroups

The *full modular group* $\Gamma_n = \mathrm{Sp}(n, \mathbb{Z})$ is the set of symplectic matrices with integral entries, in symbols,

$$\Gamma_n = \mathrm{Sp}(n, \mathbb{R}) \cap M_{2n}(\mathbb{Z}).$$

It is generated by the following symplectic matrices.

(a) $\begin{bmatrix} E & S \\ 0 & E \end{bmatrix}$, ${}^t S = S \in M_n(\mathbb{Z})$;

(b) $\begin{bmatrix} {}^t U & 0 \\ 0 & U^{-1} \end{bmatrix}$, $U \in \mathrm{GL}_n(\mathbb{Z})$ with $\det U = \pm 1$;

(c) $\begin{bmatrix} 0 & -E \\ E & 0 \end{bmatrix}$.

In particular, the full modular group Γ_n contains the following subgroup

$$\Gamma_n^0 = \left\{ \begin{bmatrix} A & B \\ C & D \end{bmatrix} \in \Gamma_n \;\middle|\; C = 0 \right\}$$

which is generated by elements of the forms in (a) and (b) as shown above. Elements in Γ_n^0 can be decomposed as

$$\begin{bmatrix} {}^tU & 0 \\ 0 & U^{-1} \end{bmatrix} \begin{bmatrix} E & S \\ 0 & E \end{bmatrix},$$

where $U \in \mathrm{GL}_n(\mathbb{Z})$ with $\det U = \pm 1$ and ${}^tS = S \in M_n(\mathbb{Z})$. Indeed Γ_n^0 is a semidirect product of its two subgroups

$$\mathcal{B} = \left\{ \begin{bmatrix} {}^tU & 0 \\ 0 & U^{-1} \end{bmatrix} \,\middle|\, U \in \mathrm{GL}_n(\mathbb{Z}), \det U = \pm 1 \right\}$$

and

$$\mathcal{P} = \left\{ \begin{bmatrix} E & S \\ 0 & E \end{bmatrix} \,\middle|\, {}^tS = S \in M_n(\mathbb{Z}) \right\}.$$

Proposition 8.1.1. *Suppose that M_1, $M_2 \in \mathrm{Sp}(n, \mathbb{Z})$ have the same C, D blocks. Then*

$$\Gamma_n^0 M_1 = \Gamma_n^0 M_2.$$

Proof. Suppose that

$$M_1 = \begin{bmatrix} * & * \\ C & D \end{bmatrix} \quad \text{and} \quad M_2 = \begin{bmatrix} * & * \\ C & D \end{bmatrix}.$$

Then

$$M_1 M_2^{-1} = \begin{bmatrix} * & * \\ C & D \end{bmatrix} \begin{bmatrix} {}^tD & * \\ -{}^tC & * \end{bmatrix} = \begin{bmatrix} * & * \\ 0 & * \end{bmatrix} \in \Gamma_n^0.$$

So it follows that $\Gamma_n^0 M_1 = \Gamma_n^0 M_2$.   □

Now we consider the converse.

Proposition 8.1.2. *Suppose that M_1, $M_2 \in \mathrm{Sp}(n, \mathbb{Z})$ such that $\Gamma_n^0 M_1 = \Gamma_n^0 M_2$ and*

$$M_2 = \begin{bmatrix} A & B \\ C & D \end{bmatrix}.$$

Then the C, D blocks of M_1 are UC, UD for some $U \in \mathrm{GL}_n(\mathbb{Z})$ with $\det U = \pm 1$.

Proof. It follows from

$$M_1 = \begin{bmatrix} {}^tU^{-1} & 0 \\ 0 & U \end{bmatrix} \begin{bmatrix} E & S \\ 0 & E \end{bmatrix} \begin{bmatrix} A & B \\ C & D \end{bmatrix} = \begin{bmatrix} * & * \\ UC & UD \end{bmatrix}.$$ □

A pair of $n \times n$ integral matrices (C, D) is called a *coprime pair* if it satisfies the following conditions:

(a) $C^t D = D^t C$;

(b) There exist two integral matrices P and Q such that the matrix

$$U = \begin{bmatrix} P & Q \\ C & D \end{bmatrix}$$

is unimodular, i.e., $\det U = \pm 1$.

Proposition 8.1.3. *For any coprime pair (C, D), there exist A, $B \in M_n(\mathbb{Z})$ such that*

$$\begin{bmatrix} A & B \\ C & D \end{bmatrix} \in \mathrm{Sp}(n, \mathbb{Z}).$$

Proof. Let U be a unimodular integral matrix with C and D as lower half blocks. From $UU^{-1} = E_{2n}$, we get integral matrices X and Y such that

$$CX + DY = E.$$

Now let $A = {}^tY + {}^tXYC$ and $B = -{}^tX + {}^tXYD$. Then

$$A^t B = B^t A$$

and $A^t D - B^t C = E$. In other words, we have

$$\begin{bmatrix} A & B \\ C & D \end{bmatrix} \in \mathrm{Sp}(n, \mathbb{Z}).$$ □

For each positive integer N, the *principal congruence subgroup of level N of Γ_n* is defined as

$$\Gamma_n(N) = \{M \in \Gamma_n \mid M \equiv E_{2n} \pmod{N}\}.$$

It is a normal subgroup of Γ_n of finite index. Indeed, it is the kernel of the natural projection from $\mathrm{Sp}(n, \mathbb{Z})$ onto $\mathrm{Sp}(n, \mathbb{Z}/N\mathbb{Z})$ and hence

$$[\Gamma_n : \Gamma_n(N)] = |\mathrm{Sp}(n, \mathbb{Z}/N\mathbb{Z})|$$
$$= N^{n(2n+1)} \prod_{p|N} (1 - p^{-2})(1 - p^{-4}) \cdots (1 - p^{-2n}).$$

8.2. Regular Elliptic Elements in the Full Modular Group

Let
$$\Gamma_n = \mathrm{Sp}(n, \mathbb{Z}) = \mathrm{Sp}(n, \mathbb{R}) \cap M_{2n}(\mathbb{Z})$$

be the full modular group. An element M in Γ_n is called a *regular elliptic element* if there exists a $Z_0 \in \mathcal{H}_n$ such that Z_0 is the unique isolated fixed point of M.

Proposition 8.2.1. *If M is a regular elliptic element of Γ_n, then M is conjugate in $\mathrm{Sp}(n, \mathbb{R})$ to an element of the form*

$$\begin{bmatrix} A & B \\ -B & A \end{bmatrix}$$

with $A + Bi = \mathrm{diag}\left[e^{i\theta_1}, e^{i\theta_2}, \ldots, e^{i\theta_n} \right]$, $\theta_j \in \mathbb{R}$. Furthermore, we have

$$e^{i(\theta_j + \theta_k)} \neq 1$$

for all $1 \leq j, k \leq n$.

Proof. Suppose that $Z_0 = X_0 + iY_0$ is the fixed point of M. Set

$$M_0 = \begin{bmatrix} E & X_0 \\ 0 & E \end{bmatrix} \begin{bmatrix} Y_0^{1/2} & 0 \\ 0 & Y_0^{-1/2} \end{bmatrix}.$$

Then

$$MM_0(iE) = M_0(iE),$$

and hence

$$M_0^{-1} M M_0(iE) = iE.$$

Thus $M_0^{-1} M M_0$ lies in the isotropy subgroup of $\mathrm{Sp}(n, \mathbb{R})$ at $Z = iE$, i.e., M is conjugate in $\mathrm{Sp}(n, \mathbb{R})$ to an element of the form

$$\begin{bmatrix} A & B \\ -B & A \end{bmatrix}, \quad A + iB \in U(n).$$

Note that elements in $U(n)$ can be diagonalized. So we can assume that $A + iB$ is indeed a diagonal matrix. Also the eigenvalues of unitary matrices have absolute value 1. It follows $A + iB = \text{diag} \left[e^{i\theta_1}, e^{i\theta_2}, \ldots, e^{i\theta_n} \right]$. Under the Cayley transform

$$Z \mapsto W = (Z - iE)(Z + iE)^{-1},$$

the upper half-plane $\mathcal{H}_n$ is transformed into the generalized disc D_n and the element

$$\begin{bmatrix} A + iB & 0 \\ 0 & A - iB \end{bmatrix}$$

acts on D_n by the action

$$W \mapsto (A + iB)W(A - iB)^{-1} = \left[w_{jk} e^{i(\theta_j + \theta_k)} \right].$$

Note that the above mapping has only one fixed point $W = 0$, it follows that

$$e^{i(\theta_j + \theta_k)} \neq 1$$

for all $1 \leq j, k \leq n$. Otherwise, it has more than one fixed point beside $W = 0$. $\qquad \square$

Proposition 8.2.2. *If M is a regular elliptic element of $\Gamma_n = \text{Sp}(n, \mathbb{Z})$, then the centralizer of M in Γ_n is a finite group.*

Proof. Denote by $C(M, \Gamma_n)$ the centralizer of M in Γ_n. Suppose that Z_0 is the isolated fixed point of M and $\gamma \in C(M, \Gamma_n)$. Then

$$M(\gamma(Z_0)) = \gamma(M(Z_0)) = \gamma(Z_0).$$

So $\gamma(Z_0)$ is also a fixed point of M. But Z_0 is the only fixed point of M, it forces that $\gamma(Z_0) = Z_0$. Thus γ lies in a compact subgroup of $\text{Sp}(n, \mathbb{R})$. Also $C(M, \Gamma_n)$ is discrete, so it must be a finite group. $\qquad \square$

Corollary 8.2.3. *If M is a regular elliptic element of Γ_n, then M is an element of finite order and its characteristic polynomial is a product of cyclotomic polynomials.*

In our construction of a fundamental domain with respect to $SL_2(\mathbb{Z})$, we already have the following.

Proposition 8.2.4. *Let M be a regular elliptic element of $\mathrm{SL}_2(\mathbb{Z})$, then M is conjugate in $\mathrm{SL}_2(\mathbb{Z})$ to a power of J or JT^{-1} with*

$$J = \begin{bmatrix} 0 & -1 \\ 1 & 0 \end{bmatrix} \quad and \quad T = \begin{bmatrix} 1 & 1 \\ 0 & 1 \end{bmatrix}.$$

In particular, M is an element of order 2 or 3 in the modular group with the characteristic polynomial $X^2 + 1$ or $X^2 \pm X + 1$.

Remark 8.2.5. The conjugacy classes of elliptic elements of $\mathrm{Sp}(2, \mathbb{Z})$ was classified by Göschling in 1962 while Eie classified the conjugacy classes of $\mathrm{Sp}(3, \mathbb{Z})$ in 1986 in an attempt to derive the dimension formula for modular forms of degree three via Selberg trace formula. It is still an open problem to classify conjugacy classes of $\mathrm{Sp}(4, \mathbb{Z})$ or full modular groups of higher degrees.

8.3. Siegel Modular Forms

Let k be a nonnegative integer. A holomorphic function $f \colon \mathcal{H}_n \to \mathbb{C}$ is a *Siegel modular form of weight k and degree n with respect to the full modular group $\Gamma_n = \mathrm{Sp}(n, \mathbb{Z})$* if it satisfies the following conditions:

(S-1) $f(M(Z)) = \det(CZ + D)^k f(Z)$ for all $M = [\begin{smallmatrix} A & B \\ C & D \end{smallmatrix}] \in \Gamma_n$;

(S-2) If $n = 1$, $f(z)$ has a Fourier expansion of the form

$$f(z) = \sum_{m=0}^{\infty} a(m)e^{2\pi imz}.$$

Remark 8.3.1. The first condition can be replaced by the following conditions since Γ_n is well-known to be generated by elements corresponding to translation and involution:

(a) $f(Z + S) = f(Z)$ for all ${}^tS = S \in M_n(\mathbb{Z})$;

(b) $f(-Z^{-1}) = (\det Z)^k f(Z)$.

From the above two conditions, we also conclude that

(c) $f(Z[U]) = (\det U)^k f(Z)$ for all $U \in \mathrm{GL}_n(\mathbb{Z})$ with $\det U = \pm 1$.

Remark 8.3.2. If f is a Siegel modular form of weight k and $M = -E_{2n}$, then

$$f(Z) = [\det(-E_n)]^k f(Z) = (-1)^{kn} f(Z).$$

So we require that k is even when n is odd.

For a modular form of degree $n > 1$, it has a Fourier expansion automatically due to Koecher principle as follows:

Koecher Principle. *Suppose that* $f \colon \mathcal{H}_n \to \mathbb{C}$ *is a holomorphic function such that*

(a) $f(Z + S) = f(Z)$ *for all* ${}^t S = S \in M_n(\mathbb{Z})$;

(b) $f(Z[U]) = f(Z)$ *for* $U \in \mathrm{GL}_n(\mathbb{Z})$, $\det U = \pm 1$.

In the cases of $n \geq 2$, f *has a Fourier expansion*

$$f(Z) = \sum_{\substack{T \in \Lambda_n^* \\ T \geq 0}} a(T) e^{2\pi i \sigma(TZ)},$$

where Λ_n^* *is the set of* $n \times n$ *half-integral symmetric matrices and* $\sigma(X)$ *is the trace of a matrix* X.

A symmetric matrix T is half-integral if $2T \in M_n(\mathbb{Z})$ and all the diagonal entries of T are integers. Given two symmetric matrices $A = [a_{ij}]$ and $B = [b_{ij}]$, the notation

$$\sigma(AB) = \sum_{i=1}^{n} \sum_{j=1}^{n} a_{ij} b_{ij}$$

is the usual inner product of $n \times n$ matrices. If Λ_n is the set of $n \times n$ integral symmetric matrices, then Λ_n^* is the dual lattice of Λ_n with respect to the above inner product.

There are two main sources of examples of Siegel modular forms, i.e., Eisenstein series and theta series. First we introduce Eisenstein series. The theta series will be studied in Section 9.3. For each

$$M = \begin{bmatrix} A & B \\ C & D \end{bmatrix} \in \mathrm{Sp}(n, \mathbb{R}),$$

we let
$$j(M, Z) = \det(CZ + D).$$

The Jacobian determinant of the transform $Z \mapsto M(Z) = (AZ+B)(CZ+D)^{-1}$ from $\mathcal{H}_n$ to $\mathcal{H}_n$ is given by $|\det(CZ+D)|^{-2(n+1)}$. As $j(M, Z) = \det(CZ+D)$ is a factor of $\det(CZ+D)^{-2(n+1)}$, the chain rule on differentiation then implies the *cocycle condition*

$$j(M_1 M_2, Z) = j(M_1, M_2(Z)) j(M_2, Z).$$

Also note that for two matrices M_1, $M_2 \in \Gamma_n$, we have

$$j(M_1, Z) = j(M_2, Z)$$

if and only if $\Gamma_n^0 M_1 = \Gamma_n^0 M_2$.

The *Eisenstein series* $E_k(Z)$ is defined as

$$E_k(Z) = \sum_{M : \Gamma_n / \Gamma_n^0} j(M, Z)^{-k}.$$

This series is absolutely convergent for $k > (n+1)$ and hence it defines a holomorphic function since the general term $j(M, Z)^{-k}$ is holomorphic. Also for $M \in \mathrm{Sp}(n, \mathbb{Z})$, we have

$$E_k(M(Z)) = \sum_{M' : \Gamma_n / \Gamma_n^0} j(M', M(Z))^{-k}$$

$$= j(M, Z)^k \sum_{M' : \Gamma_n / \Gamma_n^0} j(M'M, Z)^{-k}$$

$$= j(M, Z)^k E_k(Z).$$

It follows that $E_k(Z)$ is indeed a modular form of weight k if $k > n+1$ and $n \geq 2$.

8.4. Exercises

1. If M is a regular elliptic element of $\mathrm{Sp}(2, \mathbb{Z})$, show that the possible factors in $\mathbb{Z}[X]$ of the characteristic polynomial of M are

$$X^2 + 1, \quad X^2 \pm X + 1, \quad X^4 + X^3 + X^2 + X + 1,$$
$$X^4 - X^2 + 1, \quad X^4 + X^2 + 1, \quad X^4 + 1.$$

Also find the possible order of M.

2. Classify the conjugacy classes of regular elliptic elements in $\mathrm{Sp}(2, \mathbb{Z})$ under the conjugation of $\mathrm{Sp}(2, \mathbb{R})$. Represent the conjugacy classes by elements of the form diag $\left[e^{i\theta_1}, e^{i\theta_2}\right]$ in $U(2)$.

3. For a holomorphic function f satisfying

$$\int_{\mathcal{H}_n} (\det Y)^{k-(n+1)} |f(Z)| \, dX dY < +\infty, \quad k > (n+1),$$

and

$$\int_{\mathcal{H}_n} (\det Y)^{k-(n+1)} |f(Z)|^2 \, dX dY < +\infty, \quad k > (n+1).$$

Show that the Poincaré series

$$f_\Gamma(Z) = \sum_{M \in \Gamma_n} f(M(Z)) j(M, Z)^{-k}$$

is a modular form of weight k.

Chapter 9

The Fourier Coefficients of Eisenstein Series

The Eisenstein series $E_k(Z)$ defined by

$$E_k(Z) = \sum_{M:\Gamma_n/\Gamma_n^0} j(M, Z)^{-k}$$

is the most important example of modular forms of several variables. How to compute the Fourier coefficients of $E_k(Z)$ is the central problem in the development of theory of modular forms of several variables.

9.1. A Transformation Formula

For $\operatorname{Re} s > (n+1)/2$ and $Z \in \mathcal{H}_n$, we consider the function on $\mathbb{R}^{n(n+1)/2}$ as the set of $n \times n$ symmetric matrices over $\mathbb{R}$. Define

$$\varphi(T) = \begin{cases} (\det T)^{s-(n+1)/2} e^{2\pi i \sigma(TZ)}, & \text{if } T > 0; \\ 0, & \text{otherwise.} \end{cases}$$

Then $\varphi(T)$ is a smooth function of s when $\operatorname{Re} s$ is sufficiently large.

Let Λ_n be the set of $n \times n$ integral symmetric matrices and Λ_n^* be its

155

dual lattice with the inner product

$$\sigma(AB) = \text{trace of } AB = \sum_{i=1}^{n} a_{ii}b_{ii} + 2\sum_{i<j} a_{ij}b_{ij}.$$

Then Λ_n^* consists of $n \times n$ half-integral symmetric matrices. The well-known Poisson summation formula then implies that

$$\left(\frac{1}{2}\right)^{n(n-1)/2} \sum_{W \in \Lambda_n^*} \widehat{\varphi}(W) = \sum_{T \in \Lambda_n} \varphi(T).$$

In order to find the Fourier transform $\widehat{\varphi}$ of φ, we need the following lemmas.

Lemma 9.1.1. *For* $\operatorname{Re} s > (n-1)/2$, *we have*

$$\int_{Y>0} (\det Y)^{s-(n+1)/2} e^{-\sigma(Y)} \, dY = \pi^{n(n-1)/2} \prod_{\nu=0}^{n-1} \Gamma\left(s - \frac{\nu}{2}\right).$$

Proof. We shall prove the assertion by induction on n. Suppose that $n \geq 2$ and set

$$Y = \begin{bmatrix} Y_1 & Y_{12} \\ {}^t Y_{12} & y_n \end{bmatrix}.$$

Then

$$\det Y = (\det Y_1)(y_n - Y_1^{-1}[Y_{12}]).$$

It follows that the integral is equal to

$$\int_{Y_1>0} (\det Y_1)^{s-(n+1)/2} e^{-\sigma(Y_1)} \, dY_1 \int_D \left(y_n - Y_1^{-1}[Y_{12}]\right)^{s-(n+1)/2} e^{-y_n} \, dy_n dY_{12},$$

where D is the domain in $\mathbb{R}^n$ defined by $y_n - Y_1^{-1}[Y_{12}] > 0$. With $u = y_n - Y_1^{-1}[Y_{12}]$ as a new variable in place of y_n, the second integral is transformed into

$$\int_0^\infty u^{s-(n+1)/2} e^{-u} \, du \int_{\mathbb{R}^{n-1}} e^{-Y_1^{-1}[Y_{12}]} \, dY_{12}$$

which is equal to

$$\pi^{(n-1)/2} \Gamma\left(s - \frac{n-1}{2}\right) (\det Y_1)^{1/2}.$$

Thus our assertion follows by an induction. $\qquad\square$

Lemma 9.1.2. *For any* $G \in \mathcal{H}_n$, *we have*

$$\int_{Y>0} (\det Y)^{s-(n+1)/2} e^{2\pi i \sigma(YG)} \, dY$$

$$= [\det(-2\pi i G)]^{-s} \, \pi^{n(n-1)/4} \prod_{\nu=0}^{n-1} \Gamma\left(s - \frac{\nu}{2}\right).$$

Proof. It suffices to consider the special case for $G = iL$ with $L > 0$. Then

$$\int_{Y>0} (\det Y)^{s-(n+1)/2} e^{-2\pi\sigma(YL)} \, dY$$

$$= (\det 2\pi L)^{-s} \int_{Y>0} (\det Y)^{s-(n+1)/2} e^{-\sigma(Y)} \, dY$$

$$= (\det 2\pi L)^{-s} \pi^{n(n-1)/4} \prod_{\nu=0}^{n-1} \Gamma\left(s - \frac{\nu}{2}\right).$$

$\square$

For simplicity, we let

$$\gamma(s) = \prod_{\nu=0}^{n-1} \Gamma\left(s - \frac{\nu}{2}\right)$$

which is a special gamma function for the domain of positive definite symmetric matrices of size n.

Now we are ready to calculate the Fourier transform of φ when $Z = iY$. We have

$$\widehat{\varphi}(W) = \int_{T>0} (\det T)^{s-(n+1)/2} e^{-2\pi\sigma(TY)} e^{2\pi i \sigma(TW)} \, dT$$

$$= \int_{T>0} (\det T)^{s-(n+1)/2} e^{2\pi i \sigma(T(W+iY))} \, dT$$

$$= [\det(-2\pi i(W + iY))]^{-s} \, \pi^{n(n-1)/4} \gamma(s)$$

$$= (-2\pi i)^{-ns} [\det(W + iY)]^{-s} \, \pi^{n(n-1)/4} \gamma(s).$$

So by the Poisson summation formula, we have

$$\sum_{T \in \Lambda_n^*} (\det T)^{s-(n+1)/2} e^{2\pi i \sigma(TZ)}$$

$$= (4\pi)^{n(n-1)/4} (-2\pi i)^{-ns} \gamma(s) \sum_{\lambda \in \Lambda_n} \det(Z + \lambda)^{-s}.$$

Here we shall mention one application of the above formula. Consider the following subseries of the Eisenstein series

$$F_k(Z) = \sum_{\det C \neq 0} \det(CZ + D)^{-k} = \sum_{\det C \neq 0} (\det C)^{-k} \det(Z + C^{-1}D)^{-k},$$

where (C, D) runs over non-associated coprime pairs. Two coprime pairs (C_1, D_1) and (C_2, D_2) are associated if there exists an $U \in \mathrm{GL}_n(\mathbb{Z})$, $\det U = \pm 1$, such that

$$C_2 = UC_1 \quad \text{and} \quad D_2 = UD_1.$$

When $\det C_1 \neq 0$, we have $C_2^{-1}D_2 = C_1^{-1}D_1$.

For each rational symmetric matrix R, there exist two unimodular matrices U and V such that

$$URV = \mathrm{diag}\,[e_1, e_2, \ldots, e_n]$$

with $e_j \geq 0$, $e_{j+1} \in \mathbb{Z}e_j$. Set

$$e_j = \frac{a_j}{b_j}, \quad a_j, b_j \in \mathbb{Z}, \quad b_j > 0 \quad \text{and} \quad (a_j, b_j) = 1,$$

and define the *content* of R as

$$\nu(R) = b_1 b_2 \cdots b_n.$$

If $C_0 = \mathrm{diag}\,[b_1, b_2, \ldots, b_n]$ and $D_0 = \mathrm{diag}\,[a_1, a_2, \ldots, a_n]$, then (C_0, D_0) is a coprime pair. Let

$$\begin{bmatrix} * & * \\ C_0 & D_0 \end{bmatrix} \begin{bmatrix} U & 0 \\ 0 & V^{-1} \end{bmatrix} = \begin{bmatrix} * & * \\ C & D \end{bmatrix}.$$

Then (C, D) is a coprime pair such that

$$R = C^{-1}D \quad \text{and} \quad \nu(R) = |\det C|.$$

Suppose that

$$M_1 = \begin{bmatrix} A_1 & B_1 \\ C_1 & D_1 \end{bmatrix}, \quad M_2 = \begin{bmatrix} A_2 & B_2 \\ C_2 & D_2 \end{bmatrix} \in \Gamma_n$$

such that $C_1^{-1}D_1 = C_2^{-1}D_2$. Then

$$M_1 M_2^{-1} = \begin{bmatrix} A_1 & B_1 \\ C_1 & D_1 \end{bmatrix} \begin{bmatrix} {}^t D_2 & -{}^t B_2 \\ -{}^t C_2 & {}^t A_2 \end{bmatrix} = \begin{bmatrix} * & * \\ C_1{}^t D_2 - D_1{}^t C_2 & * \end{bmatrix}.$$

Now the relation

$$C_1^{-1}D_1 = C_2^{-1}D_2 = {}^t(C_2^{-1}D_2) = {}^tD_2{}^tC_2^{-1},$$

we get

$$C_1{}^tD_2 = D_1{}^tC_2,$$

and hence $\Gamma_n^0 M_1 = \Gamma_n^0 M_2$.

Suppose that $\left[\begin{smallmatrix} A & B \\ C & D \end{smallmatrix}\right] \in \Gamma_n$ with $\det C \neq 0$. Set $R = C^{-1}D$. Then the condition $C^t D = D^t C$ implies that R is a rational symmetric matrix. By the previous process, we can find

$$M_1 = \begin{bmatrix} A_1 & B_1 \\ C_1 & D_1 \end{bmatrix} \in \Gamma_n$$

such that $R = C_1^{-1}D_1$ with $\nu(R) = |\det C_1|$. Consequently, we have $\Gamma_n^0 M = \Gamma_n^0 M_1$ and it follows $\nu(R) = |\det C_1| = |\det C|$.

Proposition 9.1.3. *There is a one-to-one correspondence between the set of the right coset representatives of Γ_n/Γ_n^0 with $\det C \neq 0$ and the set of rational symmetric $n \times n$ matrices R under the mapping*

$$\begin{bmatrix} A & B \\ C & D \end{bmatrix} \mapsto C^{-1}D = R.$$

Furthermore, we have $\nu(R) = |\det C|$.

By the above arguments, we can rewrite $F_k(Z)$ as

$$F_k(Z) = \sum_{{}^tR=R\in M_n(\mathbb{Q})} \nu(R)^{-k} \det(Z+R)^{-k}$$

$$= \sum_{{}^tR=R\in M_n(\mathbb{Q})/\Lambda_n} \nu(R)^{-k} \sum_{\lambda\in\Lambda_n} \det(Z+R+\lambda)^{-k}.$$

So our transformation formula gives the Fourier expansion of $F_k(Z)$. The resulting is

$$F_k(Z) = \sum_{T\in\Lambda_n^*, T>0} a_k(T)e^{2\pi i\sigma(TZ)}$$

with

$$a_k(T) = C_{nk}(\det T)^{k-(n+1)/2} \sum_{{}^tR=R\in M_n(\mathbb{Q})/\Lambda_n} \nu(R)^{-k} e^{2\pi i\sigma(TR)},$$

where

$$C_{nk} = (-2\pi i)^{kn}(4\pi)^{-n(n-1)/2} \left(\prod_{\nu=0}^{n-1} \Gamma\left(k - \frac{\nu}{2}\right) \right)^{-1}.$$

The series

$$S(T) = \sideset{}{'}\sum_{{}^tR=R\in M_n(\mathbb{Q})/\Lambda_n} \nu(R)^{-k} e^{2\pi i \sigma(TR)}$$

is called the *singular series*, it has the infinite product

$$S(T) = \prod_p S_p(T)$$

with

$$S_p(T) = \sum_{{}^tR=R\in M_n(\mathbb{Q}_p)/\Lambda_n(\mathbb{Z}_p)} \nu_p(R)^{-k} e^{2\pi i \sigma(TR)}.$$

Remark 9.1.4. The calculation of Fourier coefficients of $E_k(Z) - F_k(Z)$ is reduced to the calculation of Fourier coefficients of Eisenstein series of degree less than n.

9.2. Siegel's Main Formula and Its Applications

Let S be an $m \times m$ and T be an $n \times n$ symmetric, integral and positive definite matrices. Consider the equation

$$S[X] = {}^tXSX = T, \quad X \in \mathbb{Z}^{m \times n}.$$

Denote by $A(S, T)$ the number of solutions to the above system of quadratic equations. For any positive integer q, we let $A_q(S, T)$ be the number of solutions to the congruence

$$S[X] \equiv T \pmod{q}, \quad X \in (\mathbb{Z}/q\mathbb{Z})^{m \times n}.$$

Then

$$A_q(S, T) = q^{-n(n+1)/2} \sum_{{}^tV=V\in M_n(\mathbb{Z}/q\mathbb{Z})} \sum_{X\in(\mathbb{Z}/q\mathbb{Z})^{m\times n}} e^{2\pi i(S[X]-T,V)/q}.$$

The *local density* $d_p(S,T)$ at the place p is defined as

$$d_p(S,T) = \lim_{\nu \to \infty} p^{-\nu(mn-n(n+1)/2)} A_{p^\nu}(S,T).$$

Indeed, one can express

$$p^{-\nu(mn-n(n+1)/2)} A_{p^\nu}'(S,T)$$

as a Gaussian sum and it is a constant when ν is sufficiently large. Hence the limit always exists and can be computed explicitly.

Consider a sequence of positive integers $\{q_\nu\}$ such that

(1) $q_\nu \mid q_{\nu+1}$ for all ν;

(2) there exists a ν so that $\ell \mid q_\nu$ for any given positive integer ℓ.

Siegel has shown that

$$\lim_{\nu \to \infty} q_\nu^{-(mn-n(n+1)/2)} A_{q_\nu}(S,T)$$

exists if $m > n+1$ and the limit is independent of the choice of the sequence $\{q_\nu\}$. Also by the Chinese remainder theorem, we have

$$A_{qr}(S,T) = A_q(S,T) A_r(S,T)$$

if q and r are relatively prime. Thus it follows that the limit is equal to

$$\prod_p d_p(S,T),$$

where p runs over the set of finite primes.

Now we proceed to define $A_\infty(S,T)$. Let N be a relatively compact neighborhood of T in the cone of $n \times n$ positive definite symmetric matrices as a subset of $\mathbb{R}^{n(n+1)/2}$. Define a mapping $\varphi \colon \mathbb{R}^{mn} \to \mathbb{R}^{n(n+1)/2}$ by $\varphi(X) = S[X]$. As S is positive definite, $\varphi^{-1}(N)$ has finite volume. We let

$$A_\infty(S,N) = \frac{\mathrm{vol}(\varphi^{-1}(N))}{\mathrm{vol}(N)}.$$

If N runs through a sequence of neighborhoods shrinking down to T, then the limit

$$\lim_{N \to T} A_\infty(S,N)$$

exists and it is denoted by $A_\infty(S, T)$.

Suppose that S_1 and S_2 are $m \times m$ positive definite symmetric matrices. We say that S_1 and S_2 are in the same *class* if there exists an unimodular integral matrix U such that

$$S_1[U] = S_2.$$

Besides, we call S_1 and S_2 in the same *genus* if $A_\infty(S_1, S_2)A_\infty(S_2, S_1) \neq 0$ and there exists an unimodular integral matrix U such that

$$S_1[U] \equiv S_2 \pmod{q}$$

for any positive integer q.

The number of classes in the genus of S is finite, we denote this number by h and let $S_1, S_2, \ldots, S_h$ be representatives of different classes in the genus of S. Let

$$\mu(S) = \sum_{i=1}^{h} w(S_i)^{-1},$$

where $w(S_i)$ is the order of the group of units of S_i. Then Siegel's main theorem in this case is

$$\mu(S)^{-1} \sum_{i=1}^{h} w(S_i)^{-1} A(S_i, T) = A_\infty(S, T) \prod_p d_p(S, T).$$

9.3. Theta Series

Another main resource of modular forms is the construction of theta series. For any positive definite $m \times m$ symmetric matrix S, define a *theta series* $\vartheta(S, Z)$ on the Siegel upper half-plane $\mathcal{H}_n$ by

$$\vartheta(S, Z) = \sum_{G \in \mathbb{Z}^{m \times n}} e^{\pi i(S[G], Z)}, \quad Z \in \mathcal{H}_n.$$

Here $(A, B) = \operatorname{tr}(AB)$ is the inner product of $n \times n$ symmetric matrices A and B. It is easy to see that

$$\vartheta(S, Z) = \sum_{T \geq 0} A(S, T) e^{\pi i(T, Z)}.$$

With the help of Poisson summation formula, we get the following proposition.

Proposition 9.3.1. *Let $\vartheta(S, Z)$ be the theta series defined as above. Then we have*

$$\vartheta(S, Z) = i^{mn/2}(\det S)^{-n/2}(\det Z)^{-m/2}\vartheta(S^{-1}, -Z^{-1}).$$

Proof. Let

$$f(G) = e^{-\pi(S[G], Y)}.$$

Then the Fourier transform of $f(G)$ is given by

$$\widehat{f}(W) = (\det S)^{-n/2}(\det Y)^{-m/2}e^{-\pi(S^{-1}[W], Y^{-1})}.$$

Consequently, by Poisson summation formula, we get

$$\vartheta(S, iY) = (\det S)^{-n/2}(\det Y)^{-m/2}\vartheta(S^{-1}, iY^{-1}).$$

Thus our formula holds for $Z = iY$. Both sides of the formula are analytic functions of Z, hence our assertion follows.   □

Corollary 9.3.2. *Suppose that S is an $m \times m$, integral, unimodular and even matrix, i.e., $m \equiv 0 \pmod 8$. Then $\vartheta(S, Z)$ is a modular form of weight $m/2$.*

Proof. When S is integral, unimodular and even, we have $S = S^{-1}[S]$ and hence $\vartheta(S, Z) = \vartheta(S^{-1}, Z)$. So

$$\vartheta(S, -Z^{-1}) = (\det Z)^{m/2}\vartheta(S, Z)$$

by the previous proposition. Also we have

$$\vartheta(S, Z + B) = \vartheta(S, Z)$$

for all integral symmetric matrix B. As the modular group $\Gamma_n = \mathrm{Sp}(n, \mathbb{Z})$ is generated by $\iota: Z \mapsto -Z^{-1}$ and $p_B: Z \mapsto Z + B$, it follows that $\vartheta(S, Z)$ is a modular form of weight $m/2$.   □

The connection between the Eisenstein series and the theta series is shown in the following.

Theorem 9.3.3. *Let S be any even, unimodular and positive definite symmetric matrix of size m with $m = 2k > 2(n + 1)$ and $m \equiv 0 \pmod 8$. Then*

$$E_k(Z) = \sum_{M:\Gamma_n/\Gamma_n^0} \det(CZ + D)^{-k}$$

$$= \frac{1}{\mu(S)} \sum_{i=1}^{h} w(S_i)^{-1} \vartheta(S_i, Z).$$

The Fourier coefficient $a_k(T)$ of $E_k(Z)$ is given by

$$A_\infty(S, T) \prod_p d_p(S, T).$$

Remark 9.3.4. When $m = 8\ell$, the Minkowski-Siegel formula gives the value of $\mu(S)$:

$$\mu(S) = \sum_{i=1}^{h} w(S_i)^{-1} = \left| \frac{B_{4\ell}}{8\ell} \prod_{j=1}^{4\ell-1} \frac{B_{2j}}{4j} \right|.$$

For the case $m = 8$, let S be the symmetric matrices corresponding to the quadratic form $2N\left(\sum_{j=0}^{7} g_j \alpha_j\right)$ in $g_0, g_1, \ldots, g_7$, where $\{\alpha_0, \alpha_1, \ldots, \alpha_7\}$ is a basis of integral Cayley numbers. We have

$$w(S) = 2^{14} 3^5 5^2 7 \quad \text{and} \quad \mu(S) = 2^{-14} 3^{-5} 5^{-2} 7^{-1}.$$

Thus there is only one class in the genus of S.

9.4. An Example to Siegel's Main Theorem on Quadratic Forms

Consider the Diophantine equation

$$x_1^2 + x_2^2 + x_3^2 + x_4^2 = p^\tau,$$

where p is an odd prime. For each prime number q different from p, the local density at the place q is given by

$$q^{-4}\sum_{\nu=0}^{q-1}\sum_{x_1=0}^{q-1}\sum_{x_2=0}^{q-1}\sum_{x_3=0}^{q-1}\sum_{x_4=0}^{q-1}e^{2\pi i(x_1^2+x_2^2+x_3^2+x_4^2-p^\tau)\nu/q}$$

$$= 1 + q^{-4}\sum_{\nu=1}^{q-1}e^{-2\pi i(p^\tau\nu)/q}\left(\sum_{x=0}^{q-1}e^{2\pi ix^2\nu/q}\right)^4$$

$$= 1 + q^{-4}(-1)q^2$$

$$= 1 - q^{-2}.$$

Also the local density at the place 2 is given by

$$1 + 2^{-4}e^{-\pi ip^\tau}\left(\sum_{x=0}^{1}e^{\pi ix^2}\right)^4 = 1.$$

The local density at the place p is given by

$$p^{-4(\tau+1)}\sum_{\nu=0}^{p^{\tau+1}-1}e^{-2\pi ip^\tau\nu/p^{\tau+1}}\left(\sum_{x=0}^{p^{\tau+1}-1}e^{2\pi ix^2\nu/p^{\tau+1}}\right)^4.$$

When $\nu = p^j\nu'$, $(\nu',p)=1$, $1 \le j \le \tau$, the partial sum is

$$p^{-4(\tau+1)}\left(p^{\tau+1-j}-p^{\tau-j}\right)p^{4j}p^{2\tau+2-2j} = p^{-(\tau+1)}p^j(1-p^{-1}).$$

When $(\nu,p)=1$, the partial sum is

$$p^{-4(\tau+1)}(-p^\tau)p^{2(\tau+1)} = -p^{-(\tau+2)}.$$

Thus the density is

$$1 + \sum_{j=1}^{\tau}p^{-(\tau+1)}p^j(1-p^{-1}) - p^{-(\tau+2)}$$

$$= \left(1-p^{-2}\right)\left(1+p^{-1}+p^{-2}+\cdots+p^{-\tau}\right).$$

Consequently, we have

$$\prod_p d_p = \frac{4}{3}\prod_p\left(1-p^{-2}\right)\times\left(1+p^{-1}+p^{-2}+\cdots+p^{-\tau}\right)$$

$$= \frac{4}{3}\frac{1}{\zeta(2)}\times\left(1+p^{-1}+p^{-2}+\cdots+p^{-\tau}\right)$$

$$= 8\pi^{-2}\left(1+p^{-1}+p^{-2}+\cdots+p^{-\tau}\right).$$

It remains to compute the density at the infinite place. Note that the volume of the solid

$$\mathbf{x} \in \mathbb{R}^4, \quad |\mathbf{x}|^2 \leq \rho$$

is $\pi^2 \rho^2 / 2$. Thus the density is given by

$$\lim_{\epsilon \to 0} \frac{1}{\epsilon} \left\{ \frac{\pi^2}{2} (p^\tau + \epsilon)^2 - \frac{1}{2} \pi^2 p^{2\tau} \right\} = p^\tau \pi^2.$$

Finally, we have

$$d_\infty \prod_p d_p = 8 \sum_{d | p^\tau} d.$$

This is the number of solution to the Diophantine equation

$$x_1^2 + x_2^2 + x_3^2 + x_4^2 = p^\tau.$$

Remark 9.4.1. Let $A_4(n)$ be the number of integral solutions to the equation

$$x_1^2 + x_2^2 + x_3^2 + x_4^2 = n.$$

It n is odd, then

$$A_4(n) = 8 \sum_{d | n} d.$$

In general, we have

$$A_4(n) = 8 \sum_{d | n, 4 \nmid d} d.$$

Let $\left(\frac{v}{p} \right)$ be the well-known *Legendre symbol* defined by

$$\left(\frac{v}{p} \right) = \begin{cases} 1, & \text{if } v \equiv x^2 \pmod{p}; \\ -1, & \text{otherwise.} \end{cases}$$

For any odd prime p, we let

$$S = \sum_{v=1}^{p-1} \left(\frac{v}{p} \right) e^{2\pi i v / p}.$$

Proposition 9.4.2. *Let S be defined as above. Then we have*

$$S^2 = \left(\frac{-1}{p} \right) p.$$

Proof. We have

$$S^2 = \sum_{u=1}^{p-1} \sum_{v=1}^{p-1} \left(\frac{u}{p}\right)\left(\frac{v}{p}\right) e^{2\pi i(u+v)/p}.$$

Set $u = v\alpha$, then

$$S^2 = \sum_{\alpha=1}^{p-1} \left(\frac{\alpha}{p}\right) \sum_{v=1}^{p-1} e^{2\pi i v(\alpha+1)/p}.$$

When $\alpha = -1$, the inner sum is equal to $p - 1$. When $\alpha \neq p - 1$, the inner sum is equal to -1. It follows that

$$S^2 = \left(\frac{-1}{p}\right)(p-1) + \sum_{\alpha \neq p-1} \left(\frac{\alpha}{p}\right)(-1)$$

$$= \left(\frac{-1}{p}\right)(p-1) + \left(\frac{-1}{p}\right) = \left(\frac{-1}{p}\right)p.$$

$\square$

Remark 9.4.3. Indeed, we can rewrite S as

$$\sum_{x=0}^{p-1} e^{2\pi i x^2/p}.$$

Let Q be the set of quadratic residues and R be the set of quadratic non-residues. Then

$$Q \cup R = \{1, 2, \ldots, p-1\}$$

and

$$S = \sum_{v \in Q} e^{2\pi i v/p} - \sum_{v \in R} e^{2\pi i v/p} = 1 + 2\sum_{v \in Q} e^{2\pi i v/p} = \sum_{x=0}^{p-1} e^{2\pi i x^2/p}$$

since

$$\sum_{v \in Q} e^{2\pi i v/p} + \sum_{v \in R} e^{2\pi i v/p} + 1 = 0.$$

Proposition 9.4.4. *For an odd prime p and an integer v relatively prime to p, we have*

$$\sum_{x=0}^{p-1} e^{2\pi i v x^2/p} = \left(\frac{v}{p}\right) \sum_{x=0}^{p-1} e^{2\pi i x^2/p}.$$

Proof. It is obvious when $\left(\frac{v}{p}\right) = 1$. Now suppose that $\left(\frac{v}{p}\right) = -1$. Then

$$\sum_{x=0}^{p-1} e^{2\pi i v x^2/p} = 1 + 2\sum_{v\in R} e^{2\pi i v/p} = -1 - 2\sum_{v\in Q} e^{2\pi i v/p} = -\sum_{x=0}^{p-1} e^{2\pi i x^2/p}. \quad \square$$

Let p and q be odd primes. On the one hand, we have

$$S^q = S(S^2)^{(q-1)/2} = S(-1)^{(p-1)/2 \cdot (q-1)/2} p^{(q-1)/2}$$
$$\equiv S(-1)^{(p-1)(q-1)/4}\left(\frac{p}{q}\right) \quad (\text{mod } q).$$

On the other hand, we get

$$S^q \equiv \sum_{v=1}^{p-1}\left(\frac{v}{p}\right)\zeta^{vq} \quad (\text{mod } q)$$
$$\equiv \left(\frac{q}{p}\right)\sum_{v=1}^{p-1}\left(\frac{vq}{p}\right)\zeta^{vp} \quad (\text{mod } q)$$
$$\equiv \left(\frac{q}{p}\right)S \quad (\text{mod } q).$$

Hence

$$S(-1)^{(p-1)(q-1)/4}\left(\frac{p}{q}\right) \equiv \left(\frac{q}{p}\right)S \quad (\text{mod } q).$$

Multiplying by S and cancelling $\pm p$ to yield the *reciprocity law*

$$\left(\frac{q}{p}\right) = \left(\frac{p}{q}\right)(-1)^{(p-1)(q-1)/4}.$$

9.5. Exercises

1. Let $\left(\frac{v}{p}\right)$ be the well-known Legendre symbol defined by

$$\left(\frac{v}{p}\right) = \begin{cases} 1, & \text{if } v \equiv x^2 \pmod{p}; \\ -1, & \text{otherwise.} \end{cases}$$

(a) Show that $\left(\frac{v}{p}\right) \equiv v^{(p-1)/2} \pmod{p}$, and find the value of $\left(\frac{-1}{p}\right)$.

(b) Show that $\left(\frac{u}{p}\right)\left(\frac{v}{p}\right) = \left(\frac{uv}{p}\right)$.

2. Let G_m be the group consisting of $m \times m$ complex matrices, $m = p+q$, $p \geq q > 0$, such that

$$^t\overline{M}HM = H, \quad H = \begin{bmatrix} -E_p & 0 \\ 0 & E_q \end{bmatrix}.$$

Then G_m acts on the bounded symmetric domain

$$E_q - {}^t\overline{Z}Z > 0, \quad Z \in M_{p \times q}(\mathbb{C})$$

by the action $Z \mapsto (AZ + B)(CZ + D)^{-1}$.

Chapter 10

Theory of Jacobi Forms

The study of the theory of Jacobi forms began in 1985 with a textbook written by M. Eichler and D. Zagier in an attempt to extended Maaß's work on the Saito-Kurokawa conjecture [15] which asserted the existence of a lifting from modular forms of one variable of weight $2k - 2$ to Siegel modular forms of weight k and degree two.

10.1. Jacobi Forms over Real Numbers

Let k and m be a pair of nonnegative integers. A holomorphic function $\varphi \colon \mathscr{H} \times \mathbb{C} \to \mathbb{C}$ is called a *Jacobi form of weight k and index m with respect to the Jacobi group* $\mathrm{SL}_2(\mathbb{Z}) \ltimes \mathbb{Z}^2$ if it satisfies the following conditions:

(J-1) For all $\begin{bmatrix} a & b \\ c & d \end{bmatrix}$ in $\mathrm{SL}_2(\mathbb{Z})$,

$$\varphi \left(\frac{az + b}{cz + d}, \frac{w}{cz + d} \right) = (cz + d)^k \exp \left\{ 2\pi i m c w^2 / (cz + d) \right\} \varphi(z, w);$$

(J-2) For all integers λ and μ,

$$\varphi(z, w + \lambda z + \mu) = \exp \left\{ -2\pi i m (\lambda^2 z + 2\lambda w) \right\} \varphi(z, w);$$

(J-3) $\varphi(z, w)$ has a Fourier expansion of the form

$$\varphi(z, w) = \sum_{n=0}^{\infty} \sum_{r^2 \leq 4mn} \alpha(n, r) e^{2\pi i (nz + rw)}.$$

A main source of Jacobi forms come from the coefficients of Fourier-Jacobi expansion of Siegel modular forms of degree two. Let

$$f(Z) = \sum_{T \geq 0} a(T) e^{2\pi i \sigma(TZ)}$$

be a modular form of weight k on the upper half-plane

$$\mathcal{H}_2 = \left\{ Z = \begin{bmatrix} z & w \\ w & z^* \end{bmatrix} \ \middle| \ \mathrm{Im}\, Z > 0 \right\}.$$

Set

$$T = \begin{bmatrix} n & t/2 \\ t/2 & m \end{bmatrix}$$

with n, t, m being integers. Then

$$\sigma(TZ) = nz + tw + mz^*$$

and hence we can rearrange $f(Z)$ as

$$\sum_{m=0}^{\infty} \varphi_m(z, w) e^{2\pi i m z^*},$$

where

$$\varphi_m(z, w) = \sum_{n=0}^{\infty} \sum_{t^2 \leq 4mn} \alpha\left(\begin{bmatrix} n & t/2 \\ t/2 & m \end{bmatrix}\right) e^{2\pi i (nz + tw)}.$$

Proposition 10.1.1. *Notation as above, then $\varphi(z, w)$ is a Jacobi form of weight k and index m on $\mathcal{H} \times \mathbb{C}$.*

Proof. It suffices to prove that $\varphi(z, w)$ satisfies the conditions (J-1) and (J-2). The modular group $\mathrm{SL}_2(\mathbb{Z})$ can be embedded in $\mathrm{Sp}(2, \mathbb{Z})$ through the embedding

$$\begin{bmatrix} a & b \\ c & d \end{bmatrix} \hookrightarrow \begin{bmatrix} a & 0 & b & 0 \\ 0 & 1 & 0 & 0 \\ c & 0 & d & 0 \\ 0 & 0 & 0 & 1 \end{bmatrix}.$$

Let M be such an element. Then a direct calculation shows

$$M(Z) = \begin{bmatrix} \frac{az+b}{cz+d} & \frac{w}{cz+d} \\ \frac{w}{cz+d} & z^* - \frac{cw^2}{cz+d} \end{bmatrix}.$$

From the formula

$$f(M(Z)) = (cz + d)^k f(Z),$$

we get

$$\sum_{m=0}^{\infty} \varphi_m \left(\frac{az+b}{cz+d}, \frac{w}{cz+d} \right) e^{2\pi i m(z^* - cw^2)/(cz+d)}$$

$$= (cz + d)^k \sum_{m=0}^{\infty} \varphi_m(z, w) e^{2\pi i m z^*}.$$

This implies (J-1) if we compare the coefficients of $e^{2\pi i m z^*}$.

On the other hand, for each pair of integers λ and μ, set

$$U = \begin{bmatrix} 1 & \lambda \\ 0 & 1 \end{bmatrix} \quad \text{and} \quad B = \begin{bmatrix} 0 & \mu \\ \mu & 0 \end{bmatrix}.$$

Then

$$Z[U] + B = \begin{bmatrix} z & w + \lambda z + \mu \\ * & z^* + 2\lambda w + \lambda^2 z \end{bmatrix}.$$

From

$$f(Z[U] + B) = f(Z),$$

we get the identity (J-2).   □

Here we mention another example of Jacobi forms. Let $Q: \mathbb{Z}^N \to \mathbb{Z}$ be a positive definite integral-valued quadratic form and B be the associated bilinear form, i.e.,

$$Q(x + y) = Q(x) + B(x, y) + Q(y).$$

Then for any vector $g_0 \in \mathbb{Z}^N$, the theta series

$$\vartheta_{g_0}(z, w) = \sum_{g \in \mathbb{Z}^N} e^{2\pi i (Q(g)z + B(g, g_0)w)}$$

is a Jacobi form of weight $N/2$ and index $Q(g_0)$ with respect to a certain congruence subgroup of $SL_2(\mathbb{Z})$.

10.2. Cayley Numbers

Let $\mathfrak{f}$ be a field. The Cayley numbers $\mathscr{C}_{\mathfrak{f}}$ over $\mathfrak{f}$ is an eight-dimensional vector space over $\mathfrak{f}$ with a standard basis $e_0, e_1, e_2, e_3, e_4, e_5, e_6, e_7$ satisfying the following rules of multiplication:

(i) $x e_0 = e_0 x = x$ for all $x \in \mathscr{C}_{\mathfrak{f}}$,

(ii) $e_j^2 = -e_0$, $j = 1, 2, \ldots, 7$, and

(iii) $e_1 e_2 e_4 = e_2 e_3 e_5 = e_3 e_4 e_6 = e_4 e_5 e_7 = e_5 e_6 e_1 = e_6 e_7 e_2 = e_7 e_1 e_3 = -e_0$.

For $x = \sum_{j=0}^{7} x_j e_j$ and $y = \sum_{j=0}^{7} y_j e_j$ in $\mathscr{C}_{\mathfrak{f}}$, we define the following operations on $\mathscr{C}_{\mathfrak{f}}$.

(1) Involution: $x \mapsto \overline{x} = x_0 e_0 - \sum_{j=1}^{7} x_j e_j$;

(2) Trace operator: $T(x) = x + \overline{x} = 2 x_0$;

(3) Norm operator: $N(x) = x \overline{x} = \overline{x} x = \sum_{j=0}^{7} x_j^2$;

(4) Inner product: $\sigma \colon \mathscr{C}_{\mathfrak{f}} \times \mathscr{C}_{\mathfrak{f}} \to \mathfrak{f}$

$$\sigma(x, y) = T(x\overline{y}) = T(y\overline{x}) = 2 \sum_{j=0}^{7} x_j y_j.$$

Note that we have the following further property.

$$N(x + y) = N(x) + N(y) + \sigma(x, y).$$

Denote by $\mathcal{O}$ the $\mathbb{Z}$-module in $\mathscr{C}_{\mathbb{Q}}$, generated by α_j, $j = 0, 1, 2, \ldots, 7$, as follows:

$$\alpha_0 = e_0, \quad \alpha_1 = e_1, \quad \alpha_2 = e_2, \quad \alpha_3 = -e_4,$$

$$\alpha_4 = \frac{1}{2}(e_1 + e_2 + e_3 - e_4), \quad \alpha_5 = \frac{1}{2}(-e_0 - e_1 - e_4 + e_5),$$

$$\alpha_6 = \frac{1}{2}(-e_0 + e_1 - e_2 + e_6), \quad \alpha_7 = \frac{1}{2}(-e_0 + e_2 + e_4 + e_7).$$

Elements in $\mathcal{O}$ are referred as integral Cayley numbers. This module satisfies the following conditions:

(1) $N(x) \in \mathbb{Z}$ and $T(x) \in \mathbb{Z}$ for each x in the set;

(2) The set is closed under subtraction and multiplication;

(3) The set contains 1.

Also $\mathcal{O}$ is the maximal module with the above properties.

For z in the upper half-plane

$$\mathcal{H} = \{z = x + iy \mid x, y \in \mathbb{R}, y > 0\},$$

we define the theta series ϑ as

$$\vartheta(z) = \sum_{t \in \mathcal{O}} e^{2\pi i N(t)z}.$$

In the following, we shall prove that ϑ is a modular form of weight 4.

Lemma 10.2.1. *Let α_j, $j = 0, 1, 2, \ldots, 7$ be the basis of $\mathcal{O}$ and S be the matrix corresponding to the quadratic form in g_j, $j = 0, 1, 2, \ldots, 7$, of $2N(\sum_{j=0}^{7} g_j \alpha_j)$. Then*

$$S = \begin{bmatrix} 2E_4 & B \\ {}^t B & 2E_4 \end{bmatrix} \quad with \quad B = \begin{bmatrix} 0 & -1 & -1 & -1 \\ 1 & -1 & 1 & 0 \\ 1 & 0 & -1 & 1 \\ 1 & 1 & 0 & -1 \end{bmatrix}.$$

In particular, S is unimodular, i.e., $\det S = 1$.

Proof. Here we only prove $\det S = 1$, the remaining follows as a direct computation. Note that

$$\begin{bmatrix} 2E_4 & B \\ {}^t B & 2E_4 \end{bmatrix} \begin{bmatrix} E_4 & -\frac{1}{2}B \\ 0 & E_4 \end{bmatrix} = \begin{bmatrix} 2E_4 & 0 \\ {}^t B & 2E_4 - \frac{1}{2}{}^t B B \end{bmatrix}.$$

Hence

$$\det S = \det(2E_4) \det\left(2E_4 - \frac{1}{2}{}^t B B\right) = 2^4 2^{-4} = 1. \qquad \square$$

Proposition 10.2.2. *The series ϑ is a modular form of weight 4.*

Proof. As $N(t) \in \mathbb{Z}$ for all $t \in \mathcal{O}$, we have

$$\vartheta(z+1) = \sum_{t \in \mathcal{O}} e^{2\pi i N(t)(z+1)} = \sum_{t \in \mathcal{O}} e^{2\pi i N(t)z} = \vartheta(z).$$

Now rewrite ϑ as

$$\vartheta(z) = \sum_{g \in \mathbb{Z}^8} e^{\pi i S[g]z}$$

and

$$\vartheta(iy) = \sum_{g \in \mathbb{Z}^8} e^{-\pi S[g]y}.$$

The Fourier transform of the function

$$f(x) = e^{-\pi S[x]y}$$

is given by

$$\widehat{f}(w) = y^{-4} (\det S)^{-1/2} e^{-\pi S^{-1}[w]y^{-1}}.$$

By the well-known Poisson summation formula, we have

$$\vartheta(iy) = y^{-4} \sum_{g \in \mathbb{Z}^8} e^{-\pi S^{-1}[g]y^{-1}} = y^{-4} \sum_{g \in \mathbb{Z}^8} e^{-\pi S[g]y^{-1}} = y^{-4} \vartheta(iy^{-1}).$$

It follows that

$$\vartheta(-z^{-1}) = z^4 \vartheta(z),$$

and hence $\vartheta(z)$ is a modular form of weight 4.   □

Corollary 10.2.3. *The series $\vartheta(z)$ is in fact the normalized Eisenstein series $E_4(z)$, i.e.,*

$$\vartheta(z) = E_4(z) = 1 + 240 \sum_{n=1}^{\infty} \left(\sum_{d|n} d^3 \right) e^{2\pi i n z}.$$

Corollary 10.2.4. *The number of solutions to the equation*

$$N(t) = m, \quad t \in \mathcal{O},$$

is given by

$$240 \sum_{d|m} d^3.$$

10.3. Jacobi Forms over Cayley Numbers

The theory of Jacobi forms on $\mathscr{H} \times \mathscr{C}_{\mathbb{C}}$, the product space of the upper half-plane and Cayley numbers over the complex field $\mathbb{C}$ was initiated by Eie around 1991 in order to construct the Maaß space on the Hermitian upper half-plane of degree two over Cayley numbers. They are of particular interest since they are related to modular forms on the 27-dimensional exceptional domain developed by Baily.

With notations concerning Cayley numbers, we are ready to formulate the definition of Jacobi forms on $\mathscr{H} \times \mathscr{C}_{\mathbb{C}}$. Let k and m be a pair of nonnegative integers. A holomorphic function $\varphi \colon \mathscr{H} \times \mathscr{C}_{\mathbb{C}} \to \mathbb{C}$ is a *Jacobi form of weight k and index m with respect to* $\mathrm{SL}_2(\mathbb{Z}) \ltimes \mathcal{O}^2$ if it satisfies the following conditions:

(a) For all $\left[\begin{smallmatrix} a & b \\ c & d \end{smallmatrix} \right]$ in $\mathrm{SL}_2(\mathbb{Z})$,

$$\varphi\left(\frac{az+b}{cz+d}, \frac{w}{cz+d} \right) = (cz+d)^k \exp\left\{ 2\pi i m c N(w)/(cz+d) \right\} \varphi(z,w);$$

(b) For all λ and μ in $\mathcal{O}$,

$$\varphi(z, w + \lambda z + \mu) = \exp\left\{ -2\pi i m [N(\lambda)z + \sigma(\lambda, w)] \right\} \varphi(z,w);$$

(c) φ has a Fourier expansion of the form

$$\varphi(z,w) = \sum_{n=0}^{\infty} \sum_{\substack{t \in \mathcal{O} \\ N(t) \leq mn}} \alpha(n,t) e^{2\pi i [nz + \sigma(t,w)]}.$$

Proposition 10.3.1. *Let $\psi(z)$ be a modular form of weight $k - 4$ and*

$$\vartheta_1(z,w) = \sum_{t \in \mathcal{O}} e^{2\pi i [N(t)z + \sigma(t,w)]}.$$

Then $\varphi(z,w) = \psi(z)\vartheta_1(z,w)$ is a Jacobi form of weight k and index 1.

Proof. The assertion follows if we can prove that $\vartheta_1(z,w)$ is a Jacobi form of weight 4 and index 1. Note that for all λ, $\mu \in \mathcal{O}$,

$$
\begin{aligned}
\vartheta_1(z, w + \lambda z + \mu) &= \sum_{t \in \mathcal{O}} e^{2\pi i [N(t)z + \sigma(t, w + \lambda z + \mu)]} \\
&= \sum_{t \in \mathcal{O}} e^{2\pi i [N(t)z + \sigma(t, \lambda)z + \sigma(t, w)]} \\
&= \sum_{t \in \mathcal{O}} e^{2\pi i [N(t+\lambda)z + \sigma(t+\lambda, w)]} \cdot e^{-2\pi i [N(\lambda)z + \sigma(\lambda, w)]} \\
&= e^{-2\pi i [N(\lambda)z + \sigma(\lambda, w)]} \vartheta_1(z, w).
\end{aligned}
$$

The transformation formula for $z = iy$ and $w = iv$,

$$
\vartheta_1\left(-\frac{1}{z}, \frac{w}{z}\right) = z^4 e^{2\pi i N(w)/z} \vartheta_1(z, w)
$$

follows from the Poisson summation formula. $\square$

Proposition 10.3.2. *Let $\varphi(z, w)$ be a Jacobi form of weight k, $k \geq 4$, and index 1. Then φ has the decomposition*

$$
\varphi(z, w) = \psi(z)\vartheta_1(z, w),
$$

where $\psi(z)$ is a modular form of weight $k - 4$.

Proof. Let

$$
\varphi(z, w) = \sum_{n=0}^{\infty} \sum_{\substack{t \in \mathcal{O} \\ N(t) \leq n}} \alpha(n, t) e^{2\pi i [nz + \sigma(t, w)]}.
$$

As $n \geq N(t)$, we set $n = n' + N(t)$ in the summation and it follows

$$
\begin{aligned}
\varphi(z, w) &= \sum_{n'=0}^{\infty} \sum_{t \in \mathcal{O}} \alpha(n' + N(t), t) e^{2\pi i [n'z + N(t)z + \sigma(t, w)]} \\
&= \sum_{n'=0}^{\infty} \sum_{t \in \mathcal{O}} \alpha(n', 0) e^{2\pi i [n'z + N(t)z + \sigma(t, w)]} \\
&= \sum_{n'=0}^{\infty} \alpha(n', 0) e^{2\pi i n'z} \vartheta_1(z, w).
\end{aligned}
$$

Now it is direct calculation to verify that

$$
\psi(z) = \sum_{n'=0}^{\infty} \alpha(n', 0) e^{2\pi i n'z}
$$

is indeed a modular form of weight $k - 4$. $\square$

10.4. Jacobi Forms as Vector-Valued Modular Forms

Given a positive integer m and an integral Cayley number q in $\mathcal{O}$, we define the *theta series* $\vartheta_{m,q}(z,w)$ by

$$\vartheta_{m,q}(z,w) = \sum_{\lambda \in \mathcal{O}} e^{2\pi im[N(\lambda+q/m)z+\sigma(\lambda+q/m,w)]}$$

$$= \sum_{t \in \Lambda(q)} e^{2\pi im[N(t)z+\sigma(t,w)]},$$

where

$$\Lambda(q) = \left\{ \lambda + \frac{q}{m} \mid \lambda \in \mathcal{O} \right\}.$$

Directly from the above definition, it is easy to see that

(1) $\vartheta_{m,q}(z+1,w) = e^{2\pi iN(q)/m}\vartheta_{m,q}(z,w)$;

(2) $\vartheta_{m,q}(z,w+\lambda z+\mu) = e^{-2\pi im[N(\lambda)z+\sigma(\lambda,w)]}\vartheta_{m,q}(z,w)$;

(3) $\vartheta_{m,q_1}(z,w) = \vartheta_{m,q_2}(z,w)$ if $q_1 \equiv q_2 \pmod{m}$.

We are able to decompose a Jacobi form into a finite linear combination of a set of theta series with coefficients which are elliptic modular forms.

Proposition 10.4.1. *Let f be a Jacobi form of weight k and index m with the Fourier expansion*

$$f(z,w) = \sum_{n=0}^{\infty} \sum_{\substack{t \in \mathcal{O} \\ N(t) \leq nm}} \alpha(n,t)e^{2\pi i[nz+\sigma(t,w)]}.$$

Then f has the unique expression

$$f(z,w) = \sum_{q:\mathcal{O}/m\mathcal{O}} F_q(z)\vartheta_{m,q}(z,w),$$

where

$$F_q(z) = \sum_{n \geq N(q)/m} \alpha(n,q)e^{2\pi i[n-N(q)/m]z}.$$

Proof. Note that for all $t, \lambda \in \mathcal{O}$,

$$\alpha(n + \sigma(t, \lambda) + mN(\lambda), t + m\lambda) = \alpha(n, t).$$

In other words, $\alpha(n, t)$ depends only on $t \pmod{m}$ and $nm - N(t)$. Set $t = q + m\lambda$ with q ranging over a set of representatives of $\mathcal{O}/m\mathcal{O}$ and λ ranging over all integral Cayley numbers subject to the condition

$$N(q + m\lambda) \leq nm.$$

It follows that

$$f(z, w)$$

$$= \sum_{n \geq N(q+m\lambda)/m} \sum_{q:\mathcal{O}/m\mathcal{O}} \sum_{\lambda \in \mathcal{O}} \alpha(n, q + m\lambda)e^{2\pi i[nz + \sigma(q+m\lambda, w)]}$$

$$= \sum_{n \geq N(q+m\lambda)/m} \sum_{q:\mathcal{O}/m\mathcal{O}} \sum_{\lambda \in \mathcal{O}} \alpha(n - \sigma(q, \lambda) - mN(\lambda), q)e^{2\pi i[nz + \sigma(q+m\lambda, w)]}.$$

Let $n' = n - \sigma(q, \lambda) - mN(\lambda)$ be a new variable in place of n. Then

$$n' \geq N(q)/m \quad \Longleftrightarrow \quad n \geq N(q + m\lambda)/m,$$

and

$$f(z, w)$$

$$= \sum_{q:\mathcal{O}/m\mathcal{O}} \sum_{n' \geq N(q)/m} \alpha(n', q)e^{2\pi i[n' - N(q)/m]z} \sum_{\lambda \in \mathcal{O}} e^{2\pi im[N(\lambda + q/m)z + \sigma(\lambda + q/m, w)]}$$

$$= \sum_{q:\mathcal{O}/m\mathcal{O}} F_q(z)\vartheta_{m,q}(z, w).$$

$\square$

The vector-valued function

$$F(z) = {}^t F_q(z)_{q:\mathcal{O}/m\mathcal{O}}$$

with

$$F_q(z) = \sum_{n \geq N(q)/m} \alpha(n, q)e^{2\pi i[n - N(q)/m]z}$$

is called the *vector-valued modular form corresponding to the Jacobi form*
f. Its component $F_q(z)$ is a modular form of weight $k - 4$ with respect to
the principal congruence subgroup

$$\Gamma[m] = \{M \in \mathrm{SL}_2(\mathbb{Z}) \mid M \equiv E_2 \pmod{m}\}.$$

Consequently, we can realize the vector space of Jacobi forms of weight
k and index m as a subspace of the space

$$[A(k - 4, \Gamma[m])]^{m^8},$$

where $A(k - 4, \Gamma[m])$ is the vector space of modular forms of weight $k - 4$
with respect to $\Gamma[m]$.

Proposition 10.4.2. *Let* $\vartheta_{m,q}(z, w)$ *be the theta series as defined. Then*

$$\vartheta_{m,q}\left(-\frac{1}{z}, \frac{w}{z}\right) = \left(\frac{z}{m}\right)^4 e^{2\pi i m N(w)/z} \sum_{p:\mathcal{O}/m\mathcal{O}} e^{-2\pi i \sigma(q,p)/m} \vartheta_{m,p}(z, w).$$

Proof. It suffices to prove that the equality holds for $z = iy$ and $w = iv$,
we have

$$\vartheta_{m,q}(iy, iv) = \sum_{\lambda \in \mathcal{O}} e^{-2\pi m[N(\lambda + q/m)y + \sigma(\lambda + q/m, v)]}$$

and

$$\vartheta_{m,q}\left(iy^{-1}, \frac{v}{y}\right) = \sum_{t \in \mathcal{O}} e^{-2\pi i m[N(t + q/m)y^{-1} + \sigma(t + q/m, -iv/y)]}.$$

Let S be the matrix corresponding to the quadratic form in g_j, $j = 0, 1, \ldots, 7$,
of $2N(\sum_{j=0}^{7} g_j \alpha_j)$, i.e.,

$$S = (\sigma(\alpha_i, \alpha_j))_{0 \le i,j \le 7},$$

and let $\hat{q}$ and $\hat{v}$ be the representatives of q and v with the basis $\alpha_0, \alpha_1, \ldots, \alpha_7$.
Then

$$\vartheta_{m,q}\left(iy^{-1}, \frac{v}{y}\right) = e^{-2\pi m N(v)/y} \sum_{\hat{t} \in \mathbb{Z}^8} e^{-\pi m S[\hat{t} + \hat{q}/m - i\hat{v}]y^{-1}}.$$

By a direct calculation, the Fourier transform of the function

$$f(x) = e^{-\pi m S[x + \hat{q}/m - i\hat{v}]y^{-1}}$$

is given by

$$\widehat{f}(z) = \left(\frac{y}{m}\right)^4 e^{-2\pi i \langle \widehat{q}/m - i\widehat{v}, z\rangle} e^{-\pi y S^{-1}[z]/m}.$$

Here $\langle \alpha, \beta \rangle$ is the inner product of α and β in $\mathbb{R}^8$. Thus the Poisson summation formula implies that

$$\vartheta_{m,q}\left(iy^{-1}, \frac{v}{y}\right) = e^{-2\pi m N(v)/y} \left(\frac{y}{m}\right)^4 \cdot \sum_{g \in \mathbb{Z}^8} e^{-2\pi i \langle \widehat{q}/m - i\widehat{v}, g \rangle} e^{-\pi y S^{-1}[g]/m}$$

$$= e^{-2\pi m N(v)/y} \left(\frac{y}{m}\right)^4 \sum_{t \in \mathcal{O}} e^{-2\pi i \sigma(q/m - iv, t)} e^{-2\pi y N(t)/m}.$$

Now letting $t = p + m\lambda$ with p ranging over all coset representatives of $\mathcal{O}/m\mathcal{O}$ and λ ranging over all integral Cayley numbers, we get our assertions. ☐

Fix a set of representatives $q_1, q_2, \ldots, q_{m^8}$ of $\mathcal{O}/m\mathcal{O}$ and let

$$\Theta = {}^t(\vartheta_{m,q_1}, \vartheta_{m,q_2}, \ldots, \vartheta_{m,q_{m^8}}).$$

Proposition 10.4.3. *There is a unique group homomorphism* $\psi \colon \mathrm{SL}_2(\mathbb{Z}) \to U(m^8)$ *such that*

$$\Theta\left(\frac{az+b}{cz+d}, \frac{w}{cz+d}\right) = (cz+d)^4 e^{2\pi i m c N(w)/(cz+d)} \overline{\psi\left(\begin{bmatrix} a & b \\ c & d \end{bmatrix}\right)} \Theta(z,w)$$

for all $\begin{bmatrix} a & b \\ c & d \end{bmatrix}$. *In particular, we have for* $T = \begin{bmatrix} 1 & 1 \\ 0 & 1 \end{bmatrix}$ *and* $J = \begin{bmatrix} 0 & -1 \\ 1 & 0 \end{bmatrix}$ *that*

(a) $\psi(T) = \mathrm{diag}\left[e^{-2\pi i N(q_1)/m}, e^{-2\pi i N(q_2)/m}, \ldots, e^{-2\pi i N(q_{m^8})/m}\right]$;

(b) $\psi(J) = m^{-4}(e^{2\pi i \sigma(q_\nu, q_\mu)/m})_{\nu, \mu = 1, 2, \ldots, m^8}$.

Proof. First, $\psi(T)$ and $\psi(J)$ are determined from the definition and the previous proposition. Nota that $\mathrm{SL}_2(\mathbb{Z})$ is generated by T and J. This proves our assertion. ☐

Remark 10.4.4. Indeed, we can obtain the explicit expression of ψ from $\psi(T)$ and $\psi(J)$ by a general argument proposed by Schoeneberg in 1974, the result is

$$\psi_{q,p}\left(\begin{bmatrix} a & b \\ c & d \end{bmatrix}\right)$$

$$= (mc)^{-4} \sum_{\lambda \in \mathcal{O}/c\mathcal{O}} \exp\left\{\frac{-2\pi i}{mc}[aN(q + m\lambda) - \sigma(q + m\lambda, p) + dN(p)]\right\}.$$

for $\left[\begin{smallmatrix} a & b \\ c & d \end{smallmatrix}\right] \in SL_2(\mathbb{Z})$ with $c \neq 0$. For the details, see [20] or [46, pp. 197–199].

10.5. Examples of Jacobi Forms

As shown in Proposition 10.4.1, we are able to decompose a Jacobi form into an inner product of a vector-valued modular form and a vector-valued theta series. Here we discuss the necessary and sufficient conditions for a vector-valued modular form corresponding to a Jacobi form.

Let $q_1, q_2, \ldots, q_{m^s}$ be a set of representatives of $\mathcal{O}/m\mathcal{O}$ and

$$F(z) = {}^t(F_{q_1}(z), F_{q_2}(z), \ldots, F_{m^s}(z))$$

with

$$F_q(z) = \sum_{n \geq N(q)/m} \alpha(n, q) e^{2\pi i [n - N(q)/m] z}.$$

Proposition 10.5.1. *The following statements are equivalent.*

(a) $f(z, w) = {}^t F(z) \cdot \Theta(z, w)$ *is a Jacobi form of weight k and index m.*

(b) $F\left(\frac{az+b}{cz+d}\right) = (cz+d)^{k-4} \psi\left(\left[\begin{smallmatrix} a & b \\ c & d \end{smallmatrix}\right]\right) F(z)$ *for all* $\left[\begin{smallmatrix} a & b \\ c & d \end{smallmatrix}\right] \in SL_2(\mathbb{Z})$.

Proof. Suppose that (a) is true, we have

$${}^t F\left(\frac{az+b}{cz+d}\right) \cdot \Theta\left(\frac{az+b}{cz+d}, \frac{w}{cz+d}\right) = (cz+d)^k e^{2\pi i m c N(w)/(cz+d)} \cdot {}^t F(z) \cdot \Theta(z, w).$$

By Proposition 10.4.3, we get

$${}^t F\left(\frac{az+b}{cz+d}\right) \overline{\psi\left(\left[\begin{smallmatrix} a & b \\ c & d \end{smallmatrix}\right]\right)} \Theta(z, w) = (cz+d)^{k-4} {}^t F(z) \cdot \Theta(z, w).$$

Since the components of the vector-valued theta function $\Theta(z, w)$ are linear independent and $\psi\left(\left[\begin{smallmatrix} a & b \\ c & d \end{smallmatrix}\right]\right)$ is a unitary matrix, we conclude that

$$F\left(\frac{az+b}{cz+d}\right) = (cz+d)^{k-4} \psi\left(\left[\begin{smallmatrix} a & b \\ c & d \end{smallmatrix}\right]\right) F(z).$$

This proves that (a) implies (b). The converse is a direct verification. $\square$

In the following, we shall define the *Jacobi-Eisenstein series* $E_{k,m}(z,w;q)$ and prove that they are indeed Jacobi forms.

Given $q \in \mathcal{O}$ with $N(q) \equiv 0 \pmod{m}$, we define the Jacobi-Eisenstein series

$$
E_{k,m}(z,w;q)
$$
$$
= \sum_{(c,d)=1} (cz+d)^{-k}
$$
$$
\times \sum_{\lambda \in \Lambda(q)} \exp\left\{ 2\pi i m \left[N(\lambda)\frac{az+b}{cz+d} + \sigma\left(\lambda, \frac{w}{cz+d}\right) - \frac{cN(w)}{cz+d} \right] \right\},
$$

where

$$
\Lambda(q) = \left\{ t + \frac{q}{m} \;\middle|\; t \in \mathcal{O} \right\} \quad \text{and} \quad \begin{bmatrix} a & b \\ c & d \end{bmatrix} \in \mathrm{SL}_2(\mathbb{Z}).
$$

The series used to define $E_{k,m}$ is absolutely convergent for $k \geq 7$, hence it is holomorphic. In light of the transformation formula of the theta series, we rewrite $E_{k,m}$ as

$$
E_{k,m}(z,w;q) = \sum_{M:\Gamma/\Gamma^\infty} j(M,z)^{4-k} \sum_{p:\mathcal{O}/m\mathcal{O}} \overline{\psi_{q,p}(M)} \vartheta_{m,p}(z,w),
$$

where $\Gamma = \mathrm{SL}_2(\mathbb{Z})$ and

$$
\Gamma^\infty = \left\{ \begin{bmatrix} 1 & n \\ 0 & 1 \end{bmatrix} \;\middle|\; n \in \mathbb{Z} \right\}.
$$

Thus the p-component of the vector-valued modular form corresponding to $E_{k,m}$ is given by

$$
E_p(z;q) = \sum_{M:\Gamma/\Gamma^\infty} \overline{\psi_{q,p}(M)} j(M,z)^{4-k}.
$$

Set

$$
E_{k,m}(z;q) = {}^t(E_{q_1}(z;q), E_{q_2}(z;q), \ldots, E_{q_{m^8}}(z;q)).
$$

Proposition 10.5.2. *For all $K \in \mathrm{SL}_2(\mathbb{Z})$, we have*

$$
E_{k,m}(K(z);q) = j(K,z)^{k-4}\psi(K)E_{k,m}(z;q)
$$

if $N(q) \equiv 0 \pmod{m}$.

Proof. Consider the matrix of modular functions defined by

$$G(z) = \sum_{M:\Gamma/\Gamma^\infty} \overline{\psi(M)} j(M,z)^{4-k}, \quad k \geq 7.$$

The function G dependes on the choice of the coset representatives of Γ/Γ^∞. Indeed we have

$$\psi\left(\begin{bmatrix} a+c & b+d \\ c & d \end{bmatrix}\right) = \psi(T)\psi\left(\begin{bmatrix} a & b \\ c & d \end{bmatrix}\right).$$

However, its qth row is independent of the choice since

$$\psi(T) = \operatorname{diag}\left[e^{-2\pi i N(q_1)/m}, e^{-2\pi i N(q_2)/m}, \ldots, e^{-2\pi i N(q_{m^8})/m}\right]$$

and $N(q) \equiv 0 \pmod{m}$. Note that ${}^t E_{k,m}(z;q)$ is precisely the qth row of $G(z)$. From the group properties of $\mathrm{SL}_2(\mathbb{Z})$ and the cocycle condition of j,

$$j(M, K(z)) = j(MK, z)j(K, z)^{-1}.$$

We conclude that

$${}^t E_{k,m}(K(z); q) = j(K, z)^{k-4} E_{k,m}(z;q)\overline{\psi(K^{-1})}$$

for all $K \in \mathrm{SL}_2(\mathbb{Z})$. Since $\psi(K)$ is unitary, it follows that

$$E_{k,m}(K(z); q) = j(K, z)^{k-4}\psi(K)E_{k,m}(z;q). \qquad \square$$

Proposition 10.5.3. *For $k \geq 7$, the Fourier expansion of $E_p(z;q)$ is given by*

$$E_p(z;q) = \left[\psi_{q,p}(E) + (-1)^k \psi_{q,p}(-E)\right] + \sum_{n > N(p)/m} a(n,p)e^{2\pi i(n - N(p)/m)z}$$

with

$$a(n,p) = \frac{(-2\pi i)^{k-4}}{(k-5)!m}(n - N(p)/m)^{k-5} \sum_{\substack{1 \leq d < m|c| \\ (c,d)=1}} \overline{\psi_{q,p}\left(\begin{bmatrix} a & b \\ c & d \end{bmatrix}\right)}$$

$$\times \exp\left\{2\pi i(mn - N(p)d/mc)\right\}.$$

Proof. According to $c = 0$ or not, we have

$$E_p(z; q) = [\psi_{q,p}(E) + (-1)^k \psi_{q,p}(-E)]$$

$$+ \sum_{c \neq 0} c^{4-k} \sum_{(c,d)=1} \overline{\psi_{q,p}\left(\begin{bmatrix} a & b \\ c & d \end{bmatrix}\right)} \left(z + \frac{d}{c}\right)^{4-k}.$$

Let $d = d' + \ell mc$ with $1 \leq d' < m|c|$. Also, note that

$$\psi_{q,p}\left(\begin{bmatrix} a & b + \ell ma \\ c & d + \ell ma \end{bmatrix}\right) = \psi_{q,p}\left(\begin{bmatrix} a & b \\ c & d \end{bmatrix}\right).$$

It follows that

$$\sum_{(c,d)=1} \overline{\psi_{q,p}\left(\begin{bmatrix} a & b \\ c & d \end{bmatrix}\right)} \left(z + \frac{d}{c}\right)^{4-k}$$

$$= m^{4-k} \sum_{\substack{1 \leq d' < m|c| \\ (d',c)=1}} \overline{\psi_{q,p}\left(\begin{bmatrix} a & b' \\ c & d' \end{bmatrix}\right)} \sum_{\ell \in \mathbb{Z}} \left(\frac{z}{m} + \frac{d'}{mc} + \ell\right)^{4-k}$$

$$= \frac{m^{4-k}(-2\pi i)^{k-4}}{(k-5)!} \sum_{n=1}^{\infty} n^{k-5} e^{2\pi i n z/m} \sum_{\substack{1 \leq d' < m|c| \\ (d',c)=1}} \overline{\psi_{q,p}\left(\begin{bmatrix} a & b' \\ c & d' \end{bmatrix}\right)} e^{2\pi i n d'/mc}.$$

On the other hand, we have

$$E_p(z + 1; q) = e^{-2\pi i N(p)/m} E_p(z; q).$$

This forces the coefficient of $e^{2\pi i n z/m}$ to be zero unless

$$n + N(p) \equiv 0 \pmod{m}.$$

Let $n' = (n + N(p))/m$ be a new variable in place of n. Our assertion then follows. $\qquad \square$

From the definition of $E_{k,m}(z, w; q)$, we see that

(1) $E_{k,m}(z, w; q) = E_{k,m}(z, w; q')$ if $q \equiv q' \pmod{m}$;

(2) $E_{k,m}(z, w; -q) = E_{k,m}(z, -w; q) = (-1)^k E_{k,m}(z, w; q).$

Thus $E_{k,m}(z, w; q) = 0$ if and only if k is odd and $2q \equiv 0 \pmod{m}$.

It is well-known that the number of solutions to the congruence

$$N(q) \equiv 0 \pmod{m}, \quad q \in \mathcal{O}/m\mathcal{O}$$

is $m^7 \sum_{d|m} \varphi(d)/d^4$. Also the number of solutions to the congruences

$$2q \equiv 0 \pmod{m} \quad \text{and} \quad N(q) \equiv 0 \pmod{m},$$

where $q \in \mathcal{O}/m\mathcal{O}$, is given by

$$N_m = \begin{cases} 1, & \text{if } m \equiv 1 \pmod{2}; \\ 136, & \text{if } m \equiv 2 \pmod{4}; \\ 256, & \text{if } m \equiv 0 \pmod{4}. \end{cases}$$

Thus the number of independent Jacobi-Eisenstein series is

$$\frac{1}{2} \left(m^7 \sum_{d|m} \frac{\varphi(d)}{d^4} + (-1)^k N_m \right).$$

10.6. Exercises

1. Let $Q: \mathbb{Z}^N \to \mathbb{Z}$ be a positive definite integral quadratic form and B be the associated bilinear form,

$$Q(x + y) = Q(x) + B(x, y) + Q(y).$$

For any vector $g_0 \in \mathbb{Z}^N$, define the theta series

$$\vartheta_{g_0}(z, w) = \sum_{g \in \mathbb{Z}^N} e^{2\pi i (Q(g)z + B(g, g_0)w)}.$$

Find a sufficient condition on Q so that $\vartheta_{g_0}(z, w)$ is a Jacobi form of weight $N/2$ and index $Q(g_0)$ with respect to a certain congruence subgroup of $SL_2(\mathbb{Z})$.

2. Find all Jacobi forms of weight 8 and index 1 with respect to $SL_2(\mathbb{Z}) \ltimes \mathcal{O}^2$.

3. Find the number of solutions to the equation

$$N(t_1) + N(t_2) = m, \quad t_1, t_2 \in \mathcal{O}.$$

4. Prove that the number of solutions to the congruence

$$N(q) \equiv 0 \pmod{m}, \quad q \in \mathcal{O}/m\mathcal{O}$$

is given by

$$m^7 \sum_{d|m} \frac{\varphi(d)}{d^4},$$

where $\varphi(n)$ is the Euler phi-function.

5. Suppose that

$$E_{k,1}(z, w) = \sum_{n \geq N(t)} \alpha(n, t) e^{2\pi i(nz + \sigma(t, w))}.$$

Show that

$$\alpha(n, t) = \begin{cases} 1, & \text{if } n = N(t); \\ -\frac{2(k-4)}{B_{k-4}} \sigma_{k-5}(n - N(t)), & \text{if } n > N(t), \end{cases}$$

where $\sigma_k(m)$ is the divisor function defined on p. 12.

6. Find the values and traces of $\psi\left(\left[\begin{smallmatrix} 0 & -1 \\ 1 & 1 \end{smallmatrix}\right]\right)$ and $\psi\left(\left[\begin{smallmatrix} 0 & -1 \\ 1 & 1 \end{smallmatrix}\right]^2\right)$ from $\psi(T)$ and $\psi(J)$. Also compute the traces of $\psi(T)$ and $\psi(J)$.

7. For $\left[\begin{smallmatrix} a & b \\ c & d \end{smallmatrix}\right] \in SL_2(\mathbb{Z})$ with $c \neq 0$, show that

$$\psi_{q,p}\left(\begin{bmatrix} a & b \\ c & d \end{bmatrix}\right)$$

$$= \frac{1}{(mc)^4} \sum_{\lambda:\mathcal{O}/c\mathcal{O}} \exp\left\{-\frac{2\pi i}{mc}\left[aN(q + m\lambda) - \sigma(q + m\lambda, p) + dN(p)\right]\right\}.$$

Chapter 11

Hecke Operators and Jacobi Forms

Hecke operators map Jacobi forms of weight k and index 1 into Jacobi forms of the same weight and different indices. So they set up the one-to-one correspondence between the vector space of modular forms of weight $k - 4$ and the vector space of Maaß forms of weight k.

11.1. Hecke Operators on Jacobi Forms

Given a positive integer m, we let

$$M_0 = \begin{bmatrix} m & 0 \\ 0 & 1 \end{bmatrix}$$

and $\Gamma = \mathrm{SL}_2(\mathbb{Z})$. Suppose that

$$\Gamma M_0 \Gamma = \Gamma M_1 \cup \Gamma M_2 \cup \cdots \cup \Gamma M_p$$

is a coset decomposition of the double coset $\Gamma M_0 \Gamma$ with coset representatives $M_1, M_2, \ldots, M_p$. If $\varphi(z, w)$ is a Jacobi form of weight k and index 1, we define the *Hecke operator* $T(m)$: $\varphi(z, w) \mapsto \varphi|_{T(m)}(z, w)$ by

$$\varphi|_{T(m)}(z, w) = m^{k-1} \sum_{M} (cz + d)^{-k} \exp\{-2\pi i m N(w) c/(cz + d)\}$$

$$\times \varphi\left(\frac{az + b}{cz + d}, \frac{mw}{cz + d}\right).$$

Here in the summation $M = \left[\begin{smallmatrix} a & b \\ c & d \end{smallmatrix}\right]$ ranges over all right coset representatives of Γ in $\Gamma M_0 \Gamma$.

Proposition 11.1.1. *If $\varphi(z, w)$ is a Jacobi form of weight k and index 1, then $\varphi|_{T(m)}(z, w)$ is a Jacobi form of weight k and index m.*

Proof. Note that if $\{M_1, M_2, \ldots, M_p\}$ is a set of right coset representatives, then $\{M_1 K, M_2 K, \ldots, M_p K\}$ is also a set of right coset representatives for any $K \in \Gamma$.

Suppose $M = \left[\begin{smallmatrix} a & b \\ c & d \end{smallmatrix}\right]$ is a coset representative in $\Gamma M_0 \Gamma$ and $K = \left[\begin{smallmatrix} \alpha & \beta \\ \gamma & \delta \end{smallmatrix}\right] \in \Gamma$. Set

$$MK = \begin{bmatrix} u & v \\ p & q \end{bmatrix}.$$

Then

$$\varphi|_{T(m)}\left(\frac{\alpha z + \beta}{\gamma z + \delta}, \frac{w}{\gamma z + \delta}\right)$$

$$= m^{k-1} \sum_{M} \left[c\frac{\alpha z + \beta}{\gamma z + \delta} + d\right]^{-k} \exp\left\{-2\pi i m N(w)\frac{c}{(\gamma z + \delta)(pz + q)}\right\}$$

$$\times \varphi\left(\frac{uz + v}{pz + q}, \frac{mw}{pz + q}\right).$$

Note that

$$-\frac{c}{(\gamma z + \delta)(pz + q)} = \frac{\gamma}{\gamma z + \delta} - \frac{p}{pz + q}$$

and

$$\left[c\frac{\alpha z + \beta}{\gamma z + \delta} + d\right]^{-k} = (\gamma z + \delta)^k (pz + q)^{-k}.$$

It follows that

$$\varphi|_{T(m)}\left(\frac{\alpha z + \beta}{\gamma z + \delta}, \frac{w}{\gamma z + \delta}\right)$$

$$= m^{k-1}(\gamma z + \delta)^k \exp\left\{2\pi i m N(w)\frac{\gamma}{\gamma z + \delta}\right\}$$

$$\times \sum_{MK} (pz + q)^{-k} \exp\left\{-2\pi i m N(w)\frac{p}{pz + q}\right\} \varphi\left(\frac{uz + v}{pz + q}, \frac{mw}{pz + q}\right)$$

$$= (\gamma z + \delta)^k \exp\left\{2\pi i m N(w)\frac{\gamma}{\gamma z + \delta}\right\} \varphi|_{T(m)}(z, w).$$

In the following consideration, we choose the following particular right coset representatives

$$\begin{bmatrix} a & b \\ 0 & d \end{bmatrix}, \quad ad = m, \quad 0 \le b < d$$

for $\Gamma M_0 \Gamma$. With such a set of representatives, the Hecke operator $T(m)$ is given by

$$\varphi|_{T(m)}(z, w) = \sum_{ad=m} \sum_{0 \le b < d} \frac{m^{k-1}}{d^k} \varphi\left(\frac{az + b}{d}, \frac{mw}{d}\right).$$

Let $\lambda, \mu \in \mathcal{O}$. Then

$$\varphi|_{T(m)}(z, w + \lambda z + \mu) = \sum_{ad=m} \sum_{0 \le b < d} \frac{m^{k-1}}{d^k} \varphi\left(\frac{az + b}{d}, \frac{m(w + \lambda z + \mu)}{d}\right).$$

Put $z^* = (az + b)/d$ and $w^* = mw/d$. Since

$$\varphi\left(\frac{az + b}{d}, \frac{m(w + \lambda z + \mu)}{d}\right) = \varphi\left(z^*, w^* + d\lambda z^* + \frac{m\mu}{d} - \lambda b\right)$$

$$= \exp\left\{-2\pi i \left[N(d\lambda)z^* + \sigma(d\lambda, w^*)\right]\right\} \varphi(z^*, w^*)$$

$$= \exp\left\{-2\pi i m \left[N(\lambda)z + \sigma(\lambda, w)\right]\right\} \varphi(z^*, w^*),$$

it follows that

$$\varphi|_{T(m)}(z, w + \lambda z + \mu) = \exp\left\{-2\pi i m \left[N(\lambda)z + \sigma(\lambda, w)\right]\right\} \varphi|_{T(m)}(z, w).$$

This proves that $\varphi|_{T(m)}(z, w)$ is a Jacobi form of weight k and index m.

Finally, we compute the Fourier expansion of $\varphi|_{T(m)}(z, w)$. Suppose that

$$\varphi(z, w) = \sum_{n=0}^{\infty} \sum_{\substack{t \in \mathcal{O} \\ N(t) \le n}} \alpha(n, t) e^{2\pi i [nz + \sigma(t, w)]}.$$

Then

$$\varphi|_{T(m)}(z, w) = \sum_{n=0}^{\infty} \sum_{\substack{t \in \mathcal{O} \\ N(t) \le n}} \alpha(n, t) \sum_{ad=m} \sum_{0 \le b < d} \frac{m^{k-1}}{d^k} e^{2\pi i [n(az+b)/d + \sigma(t, mw/d)]}.$$

Note that

$$\sum_{0 \le b < d} e^{2\pi i n b/d} = \begin{cases} d, & \text{if } d \mid n; \\ 0, & \text{otherwise.} \end{cases}$$

It follows that

$$\varphi|_{T(m)}(z, w) = \sum_{n=0}^{\infty} \sum_{\substack{t \in \mathcal{O} \\ N(t) \le n}} \alpha(n, t) \sum_{d \mid m} \left(\frac{m}{d} \right)^{k-1} e^{2\pi i [mnz/d^2 + \sigma(mt/d, w)]}.$$

So the coefficient of $e^{2\pi i [nz + \sigma(t, w)]}$ is give by

$$\sum_{d \mid (m, n, t)} d^{k-1} \alpha \left(\frac{mn}{d^2}, \frac{t}{d} \right). \qquad \square$$

11.2. Maaß Space

Let $\mathcal{H}_2$ be the domain represented by 2×2 Hermitian matrices over Cayley numbers with positive definite imaginary part. More precisely, we have

$$\mathcal{H}_2 = \left\{ Z = \begin{bmatrix} z & w \\ \overline{w} & z^* \end{bmatrix} \middle| z, z^* \in \mathbb{C}, w \in \mathscr{C}_{\mathbb{C}}, \text{Im } Z > 0 \right\}.$$

Let Γ_2 be the arithmetic subgroup of bi-holomorphic mappings from $\mathcal{H}_2$ into itself, generated by the following transformations:

(i) $P_B \colon Z \mapsto Z + B$, $B = \begin{bmatrix} n & t \\ \overline{t} & m \end{bmatrix}$, $n, m \in \mathbb{Z}, t \in \mathcal{O}$;

(ii) $T_U \colon Z \mapsto Z[U]$, $U = \begin{bmatrix} 0 & 1 \\ -1 & 0 \end{bmatrix}$ or $\begin{bmatrix} 1 & t \\ 0 & 1 \end{bmatrix}$, $t \in \mathcal{O}$;

(iii) $\iota \colon Z \mapsto -Z^{-1}$.

A holomorphic function f defined on $\mathcal{H}_2$ is a *modular form of weight k with respect to* Γ_2 if it satisfies the following conditions:

(a) $f(Z[U] + B) = f(Z)$, B, U as given in (i) and (ii);

(b) $f(-Z^{-1}) = (\det Z)^k f(Z)$.

In particular, we have $f(Z + B) = f(Z)$ and hence f has a Fourier expansion

$$f(Z) = \sum_{\substack{n,m \geq 0 \\ nm \geq N(t)}} \sum_{t \in \mathcal{O}} \alpha_f \begin{pmatrix} n & t \\ \bar{t} & m \end{pmatrix} e^{2\pi i[nz + \sigma(t,w) + mz^*]}.$$

A modular form f belongs to the Maaß space $\mathcal{M}(k, \mathscr{C})$ if its coefficients satisfy the so-called *Maaß condition* as follows:

$$\alpha_f \begin{pmatrix} n & t \\ \bar{t} & m \end{pmatrix} = \sum_{d | (m,n,t)} d^{k-1} \alpha_f \begin{pmatrix} mn/d^2 & t/d \\ \bar{t}/d & 1 \end{pmatrix}.$$

Here we give a modular form of weight 4 on $\mathcal{H}_2$. Let

$$f_4(Z) = \sum_{h \in \mathcal{O}^2} e^{2\pi i(h^t \bar{h}, Z)}, \quad Z \in \mathcal{H}_2.$$

Here the product (T, Z) is given by

$$\mathrm{tr}(TZ) = nz + \sigma(t,w) + mz^*$$

if $T = \begin{bmatrix} n & t \\ \bar{t} & m \end{bmatrix}$ and $Z = \begin{bmatrix} z & w \\ w & z^* \end{bmatrix}$.

Observe that $f_4(Z)$ is a singular modular form of weight 4 and its Fourier coefficients satisfy the following conditions:

(1) $\alpha(T) = 0$ unless $T = h^t \bar{h}$ for some $h \in \mathcal{O}^2$;

(2) $\alpha \begin{pmatrix} n & 0 \\ 0 & 0 \end{pmatrix} = 240 \sum_{d|n} d^3$ if n is a positive integer.

Therefore, $f_4(Z)$ is a Maaß form.

Proposition 11.2.1. *Suppose that $f(Z)$ is a Maaß form of weight k with $k \geq 4$. Then*

$$\varphi_0(z) = \sum_{n=0}^{\infty} \alpha_f \begin{pmatrix} n & 0 \\ 0 & 0 \end{pmatrix} e^{2\pi inz}$$

is a constant multiple of $E_k(z)$. More precisely, we have

$$\varphi_0(z) = \left(-\frac{B_k}{2k} \right) E_k(z) \alpha_f \begin{pmatrix} 1 & 0 \\ 0 & 0 \end{pmatrix}.$$

Proof. For each positive integer n, we have

$$\alpha_f \begin{pmatrix} n & 0 \\ 0 & 0 \end{pmatrix} = \sum_{d|n} d^{k-1} \alpha_f \begin{pmatrix} 1 & 0 \\ 0 & 0 \end{pmatrix}.$$

On the other hand, we have

$$E_k(z) = 1 - \frac{2k}{B_k} \sum_{n=1}^{\infty} \left(\sum_{d|n} d^{k-1} \right) e^{2\pi i n z}.$$

Hence if $\varphi_0(z) \neq 0$, then

$$\varphi_0(z) = \left(-\frac{B_k}{2k} \right) E_k(z) \alpha_f \begin{pmatrix} 1 & 0 \\ 0 & 0 \end{pmatrix}. \qquad \square$$

Proposition 11.2.2. *Suppose $f(Z)$ is a Maaß form of weight k with the expansion*

$$f(Z) = \sum_{m=0}^{\infty} \varphi_m(z, w) e^{2\pi i m z^*}.$$

Then

$$\varphi_1|_{T(m)}(z, w) = \varphi_m(z, w).$$

Proof. The coefficient of $e^{2\pi i [nz + \sigma(t, w)]}$ of the Fourier expansion of $\varphi_1|_{T(m)}$ is given by

$$\sum_{d|(m,n,t)} \alpha \begin{pmatrix} mn/d^2 & t/d \\ \bar{t}/d & 1 \end{pmatrix} d^{k-1},$$

which is precisely the Fourier coefficient $\alpha \begin{pmatrix} n & t \\ \bar{t} & m \end{pmatrix}$ of $\varphi_m(z, w)$. $\qquad \square$

Let $A_k(\Gamma)$ be the vector space of modular forms of weight k on $\mathscr{H}$ with respect to Γ. For even integer $k \geq 4$, we define a correspondence between $\mathcal{M}(k, \mathscr{C})$ and $A_{k-4}(\Gamma)$ by

$$f \rightsquigarrow \psi_1(z) = \sum_{m=0}^{\infty} \alpha_f \begin{pmatrix} n & 0 \\ 0 & 1 \end{pmatrix} e^{2\pi i n z}.$$

This is a one-to-one correspondence as shown by previous propositions. Indeed, given any modular form $g(z)$ in $A_{k-4}(\Gamma)$, if we let

$$\varphi_1(z, w) = g(z) \vartheta_1(z, w),$$
$$\varphi_m(z, w) = \varphi_1|_{T(m)}(z, w)$$

and

$$\varphi_0(z) = \left(-\frac{B_k}{2k}\right) E_k(z) a_0,$$

where a_0 is the constant term of $g(z)$. Then

$$f(Z) = \sum_{m=0}^{\infty} \varphi_m(z, w) e^{2\pi i m z^*}$$

is a Maaß form of weight k on $\mathcal{H}_2$.

11.3. Selberg Trace Formula

Recall that for any modular form φ of weight k with respect to $\Gamma[m]$, one has

$$\delta_m \varphi(z) = \frac{k-1}{4\pi} \int_{\Gamma[m] \backslash \mathcal{H}} (y')^k \sum_{M \in \Gamma[m]} \left\{ H(z, M(z')) \overline{j(M, z')} \right\}^{-k}$$
$$\times \varphi(z') \, d\mu(z'),$$

where $\delta_m = 1$ or 2 depends on $\Gamma[m]$ containing $-E_2$ or not, $H(z, z') = (z - \overline{z'})/2i$ and $d\mu(z') = dx' dy'/y'^2$.

Note that each component of the vector-valued modular form corresponding to a Jacobi form is a modular form with respect to $\Gamma[m]$. Thus we have the following proposition.

Proposition 11.3.1. *Let $F(z)$ be the vector-valued modular form corresponding to a Jacobi form f of weight k and index m. Then for $k \geq 7$, one has*

$$F(z) = \frac{k-5}{8\pi} \int_{\Gamma \backslash \mathcal{H}} (y')^{k-4} \sum_{M \in \Gamma} \left\{ H(z, M(z')) \overline{j(M, z')} \right\}^{4-k}$$
$$\times \psi(M) F(z') \, d\mu(z').$$

Let $J_{k,m}(\Gamma)$ be the vector space of Jacobi forms of weight k and index m with respect to $\Gamma = \mathrm{SL}_2(\mathbb{Z})$. A Jacobi form f is a Jacobi cusp form if f has a Fourier expansion of the form

$$\sum_{n > N(t)/m} \sum_{t \in \mathcal{O}} \alpha(n, t) e^{2\pi i [nz + \sigma(t, w)]}.$$

So the vector-valued modular form corresponding to f is a vector-valued cusp form. Denote by $J^0_{k,m}(\Gamma)$ the vector space of Jacobi cusp forms. We then have

$$\dim J_{k,m}(\Gamma) = \dim J^0_{k,m}(\Gamma) + \text{ numbers of independent Eisenstein series}$$

$$= \dim J^0_{k,m}(\Gamma) + \frac{1}{2}\left\{ m^7 \sum_{d|m} \frac{\varphi(d)}{d^4} + (-1)^k N_m \right\}.$$

Then a standard argument leads to Selberg trace formula for Jacobi cusp forms.

Theorem 11.3.2 (Selberg Trace Formula). *For all positive integers* $k > 4$ *and* m, *we have*

$$\dim J^0_{k,m} = \frac{k-5}{8\pi} \int_{\Gamma\backslash\mathscr{H}} y^{k-4} \sum_{M\in\Gamma} \left\{ H(z, M(z)) \overline{j(M,z)} \right\}^{4-k}$$

$$\times \operatorname{tr}(\psi(M))\, d\mu(z)$$

$$= \frac{k-5}{8\pi} \int_{\Gamma\backslash\mathscr{H}} y^{k-4} \sum_{M\in\Gamma} \left\{ \frac{1}{2i}\left(z - \overline{M(z)}\right) \overline{j(M,z)} \right\}^{4-k}$$

$$\times \operatorname{tr}(\psi(M))\, d\mu(z).$$

A direct calculation then yields

$$\dim J^0_{k,m} = \frac{k-5}{24}\left[m^8 + (-1)^k \gcd(m,2)^8 \right] + \frac{1}{8} i^{-k} \gcd(m,2)^4 \left[1 + (-1)^k \right]$$

$$- \left(\frac{k+1}{3}\right) \frac{1}{6}\left[\gcd(m,3)^4 + (-1)^k \right]$$

$$- \frac{1}{4}\left[m^7 \sum_{d|m} \frac{\varphi(d)}{d^4} + (-1)^k N_m \right],$$

where $\left(\frac{k+1}{3}\right)$ denotes the Legendre symbol.

Remark 11.3.3. The Peterson inner product for f, $g \in J_{k,m}(\Gamma)$ with at least one of f and g is a cusp form, was given as

$$\langle f, g \rangle = \int_{\mathscr{F}} f(z,w)\overline{g(z,w)} y^k e^{-4\pi m N(v)/y}\, dw,$$

where $w = u + iv$ and $dw = y^{-10}\, dx\,dy\,du\,dv$ is the invariant volume element.

11.4. Exercises

1. Let $E_{k,m}(z, w; q)$ be the Jacobi-Eisenstein series as defined in Chapter 10. Suppose that

$$E_{k,m}(z, w; q) = \sum_{m=0}^{\infty} \varphi_m(z, w)e^{2\pi i m z^*}.$$

   For $m \geq 2$, show that $\varphi_m(z, w)$ is not the image of the Hecke operator $T(m)$ at $\varphi_1(z, w)$.

2. Let $\Gamma[n]$ be the principal congruence subgroup of $\Gamma = SL_2(\mathbb{Z})$ of level n. For $m \geq 3$, find the dimension of $J_{k,m}^0(\Gamma[m])$, the vector space of Jacobi forms of weight k and and index m with respect to the group $\Gamma[m]$, via the Selberg trace formula.

3. It is well-known that the Eisenstein series

$$E_k(Z) = \sum_{M:\Gamma_2/\Gamma_2^0} j(M, Z)^{-k}$$

   is in the Maaß space. According to such a fact, find the Fourier coefficients of $E_k(Z)$.

4. Determine whether the power $[f_4(Z)]^n$ is a Maaß form or not.

Chapter 12

Singular Modular Forms on the Exceptional Domain

The exceptional domain of dimension 27 is a tube domain in the 3×3 Hermitian matrices over Cayley numbers. With the theory of Jacobi forms of degree two over Cayley numbers, we are able to construct a singular modular form of weight 4 on the exceptional domain.

12.1. Modular Forms on the Exceptional Domain

Let $\mathscr{T}_\mathbb{R}$ be the set of 3×3 Hermitian matrices over real Cayley numbers. The set $\mathscr{T}_\mathbb{R}$ consists of matrices of the following form

$$X = \begin{bmatrix} \xi_1 & x_{12} & x_{13} \\ \overline{x}_{12} & \xi_2 & x_{23} \\ \overline{x}_{13} & \overline{x}_{23} & \xi_3 \end{bmatrix}, \quad \xi_1, \xi_2, \xi_3 \in \mathbb{R}, \quad x_{12}, x_{13}, x_{23} \in \mathscr{C}_\mathbb{R}. \quad (12.1.1)$$

For $X \in \mathscr{T}_\mathbb{R}$ as given in (12.1.1), we define

(a) $\operatorname{tr}(X) = \xi_1 + \xi_2 + \xi_3$;

199

(b) $\det(X) = \xi_1\xi_2\xi_3 - \xi_1 N(x_{23}) - \xi_2 N(x_{13}) - \xi_3 N(x_{12}) + T((x_{12}x_{23})\overline{x}_{13})$;

(c) $X \times X = X^2 - \text{tr}(X)X + \frac{1}{2}(\text{tr}(X)^2 - \text{tr}(X^2))E$, or

$$\begin{bmatrix} \xi_2\xi_3 - N(x_{23}) & x_{13}\overline{x}_{23} - \xi_3 x_{12} & x_{12}x_{23} - \xi_2 x_{13} \\ x_{23}\overline{x}_{13} - \xi_3\overline{x}_{12} & \xi_1\xi_3 - N(x_{13}) & \overline{x}_{13}x_{13} - \xi_1 x_{23} \\ \overline{x}_{23}\overline{x}_{12} - \xi_2\overline{x}_{13} & \overline{x}_{13}x_{12} - \xi_1\overline{x}_{23} & \xi_1\xi_2 - N(x_{12}) \end{bmatrix}.$$

Note that X is invertible if and only if $\det X \neq 0$. In this case, the inverse is given by

$$X^{-1} = \frac{1}{\det X}(X \times X).$$

Also we set

$$\text{rank}\, X = 1 \iff X \neq 0, X \times X = 0,$$
$$\text{rank}\, X = 2 \iff X \times X \neq 0, \det X = 0,$$
$$\text{rank}\, X = 3 \iff \det X \neq 0.$$

We supply $\mathscr{T}_{\mathbb{R}}$ with a product $X \circ Y$ defined by

$$X \circ Y = \frac{1}{2}(XY + YX), \tag{12.1.2}$$

where XY is the ordinary matrix product. Then $\mathscr{T}_{\mathbb{R}}$ becomes a real Jordan algebra with this product. Define an inner product on $\mathscr{T}_{\mathbb{R}}$ by

$$(X, Y) = \text{tr}(X \circ Y). \tag{12.1.3}$$

Finally, we let $\mathcal{K}$ be the set of squares $X \circ X$ of elements of $\mathscr{T}_{\mathbb{R}}$ and $\mathcal{K}^+$ be the interior of $\mathcal{K}$. The *exceptional domain* $\mathcal{H}$ in $\mathbb{C}^{27}$ is then defined by

$$\mathcal{H} = \{Z = X + iY \mid X, Y \in \mathscr{T}_{\mathbb{R}}, Y \in \mathcal{K}^+\}. \tag{12.1.4}$$

Set $\mathscr{T}_{\mathcal{O}} = \mathscr{T}_{\mathbb{R}} \cap M_3(\mathcal{O})$. Here $M_3(\mathcal{O})$ is the set of 3×3 matrices over integral Cayley numbers. For $1 \leq i, j \leq 3$, let e_{ij} be the 3×3 matrix with 1 at the (i,j)-entry and 0 elsewhere. When $i \neq j$ and $t \in \mathscr{C}_{\mathbb{R}}$, we let $U_{ij}(t) = E + te_{ij}$, E being the 3×3 identity matrix.

The group of holomorphic automorphisms $\mathcal{G}$ of $\mathcal{H}$ is a Lie group of type E_7 [5]. Let Γ' be the discrete subgroup of $\mathcal{G}$ generated by the following automorphisms of $\mathcal{H}$.

(1) $\iota \colon Z \mapsto -Z^{-1}$;

(2) $p_B \colon Z \mapsto Z + B$, $B \in \mathscr{T}_{\mathcal{O}}$;

(3) $t_U \colon Z \mapsto {}^t\overline{U}ZU$, $U = U_{ij}(t)$, $t \in \mathcal{O}$.

Let k be an even integer. A holomorphic function f defined on $\mathcal{H}$ is a *modular form of weight k with respect to* Γ' if it satisfies the following conditions:

(a) $f(-Z^{-1}) = (\det(-Z))^k f(Z)$;

(b) $f(Z[U] + B) = f(Z)$ for all $B \in \mathscr{T}_{\mathcal{O}}$ and $U = U_{ij}(t)$, $t \in \mathcal{O}$.

In particular, from (b), a modular form f on $\mathcal{H}$ has a Fourier expansion of the form

$$f(Z) = \sum_{T \in \mathcal{K} \cap \mathscr{T}_{\mathcal{O}}} a(T) e^{2\pi i(T,Z)}.$$

We say that f is a *singular modular form* if $a(T) = 0$ unless $\det T = 0$.

Baily [5] considered the *Eisenstein series* $E_\ell(Z)$ by

$$E_\ell(Z) = \sum_{\gamma \in \Gamma'/\Gamma_0'} J(\gamma, Z)^\ell, \quad Z \in \mathcal{H}. \tag{12.1.5}$$

Here Γ_0' is the subgroup of Γ' generated by p_B and t_U with $B \in \mathscr{T}_{\mathcal{O}}$, $U = U_{ij}(t)$, $t \in \mathcal{O}$. And $J(\gamma, Z)$ is the determinant of the Jacobian matrix of γ at Z and have the following properties.

(1) $J(p_B, Z) = 1$ for all $B \in \mathscr{T}_{\mathbb{R}}$;

(2) $J(t_U, Z) = 1$ for all $U = U_{ij}(t)$, $t \in \mathcal{O}$;

(3) $J(\iota, Z) = (\det(-Z))^{-18}$.

For any positive even integer ℓ, the series in (12.1.5) converges absolutely and uniformly on any compact subset of $\mathcal{H}$. Hence E_ℓ is a modular form of weight 18ℓ with respect to Γ' on $\mathcal{H}$, and has a Fourier expansion

$$E_\ell(Z) = \sum_{T \in \mathscr{T}_{\mathcal{O}} \cap \mathcal{K}} a_\ell(T) e^{2\pi i(T,Z)}. \tag{12.1.6}$$

Baily proved that the Fourier coefficients $a_\ell(T)$ of $E_\ell(Z)$ are rational numbers and concluded that Satake compactification of $\mathcal{H}/\Gamma'$ has a biregularly equivalent projective model defined over the rational number field.

It is a long-standing problem to construct theta series on the exceptional domain since Baily initiated the study of Eisenstein series on the exceptional domain in 1970. However, in 1993, Kim was able to constructed a singular form of weight 4 on the exceptional domain by the analytic continuation of a nonholomorphic Eisenstein series. Here we describe his construction. Let $j(g, Z)$ be a factor of $J(g, Z)$ with the following properties.

(1) $j(p_B, Z) = 1$ for all $B \in \mathscr{F}_\mathbb{R}$;

(2) $j(t_U, Z) = 1$ for all $U = U_{ij}(t)$, $t \in \mathcal{O}$;

(3) $j(\iota, Z) = \det(-Z)$;

(4) $j(g_1 g_2, Z) = j(g_1, g_2(Z)) j(g_2, Z)$ for all g_1, $g_2 \in \Gamma'$.

Moreover, we have
$$J(g, Z) = j(g, Z)^{-18}.$$

A holomorphic function on the exceptional domain is a *modular form of weight k* if it satisfies

$$f(\gamma(Z)) = j(\gamma, Z)^k f(Z)$$

for all $\gamma \in \Gamma'$.

Now define an Eisenstein series as follows:

$$E_{k,s}(Z) = \sum_{\gamma \in \Gamma'/\Gamma'_0} j(\gamma, Z)^{-k} |j(\gamma, Z)|^{-s},$$

where k is a positive integer and $s \in \mathbb{C}$ such that $k + \operatorname{Re} s > 18$.

Among other things, Kim proved that $E_{k,s}$ can be continued as a meromorphic function in s to the whole complex plane. Moreover, $E_{4,0}(Z)$ and $E_{8,0}(Z)$ are singular modular forms with

$$E_{4,0}(Z) = 1 + 240 \sum_{\substack{T \in \mathscr{T}_0^+ \\ \operatorname{rank} T = 1}} \sigma_3(\epsilon(T)) e^{2\pi i(T, Z)},$$

where $\epsilon(T)$ is the largest positive integer d such that $d^{-1}T \in \mathscr{T}_\mathcal{O}$.

12.2. Jacobi Forms of Degree Two over Cayley Numbers

By a Jacobi form of degree two over Cayley numbers, we mean a Jacobi form defined on $\mathcal{H}_2 \times \mathscr{C}^2$. Let k and $m \geq 1$ be nonnegative integers. A holomorphic function $f \colon \mathcal{H}_2 \times \mathscr{C}^2 \to \mathbb{C}$ is called a *Jacobi form of weight k and index m with respect to* Γ_2 if f satisfies the following conditions:

(J-1) $f(Z + B, W) = f(Z, W)$ for $B \in \Lambda_2$;

(J-2) $f(Z[U], {}^t\overline{U}W) = f(Z, W)$ for all $U = \left[\begin{smallmatrix} 1 & 0 \\ t & 1 \end{smallmatrix}\right]$ or $\left[\begin{smallmatrix} 0 & 1 \\ -1 & 0 \end{smallmatrix}\right]$, $t \in \mathcal{O}$;

(J-3) $f(-Z^{-1}, Z^{-1}W) = (\det Z)^k \exp\left\{2\pi i m Z^{-1}[W]\right\} f(Z, W)$;

(J-4) $f(Z, W + Zq + p) = \exp\left\{-2\pi i m(Z, q^t\overline{q}) + \sigma(w_1, q_1) + \sigma(w_2, q_2)\right\} f(Z, W)$
for all $q, p \in \mathcal{O}^2$;

(J-5) f has a Fourier expansion of the form

$$f(Z, W) = \sum_{q \in \mathcal{O}^2} \sum_{\substack{T \in \Lambda_2 \\ mT \geq q^t\overline{q}}} a(T, q) e^{2\pi i[(T, Z) + \sigma(q_1, w_1) + \sigma(q_2, w_2)]}. \qquad (12.2.1)$$

Here for $Z = \left[\begin{smallmatrix} z_1 & z_{12} \\ \overline{z}_{12} & z_2 \end{smallmatrix}\right] \in \mathcal{H}_2$ and $W = \left[\begin{smallmatrix} w_1 \\ w_2 \end{smallmatrix}\right] \in \mathscr{C}_{\mathbb{C}}^2$, we let

$$Z[W] = z_1 N(w_1) + z_2 N(w_2) + \sigma(z_{12}w_2, w_1).$$

For 2×2 Hermitian matrix $A = \left[\begin{smallmatrix} a & \lambda \\ \overline{\lambda} & b \end{smallmatrix}\right]$, we write $A \geq 0$ to mean $a \geq 0$, $ab \geq N(\lambda)$. Also $A \geq B$ if and only if $A - B \geq 0$.

From the above definition, we are able to decompose a Jacobi form of degree two into an inner product of a vector-valued modular form and a vector theta series.

Proposition 12.2.1. *Let $f(Z, W)$ be a Jacobi form of degree two with Fourier expansion (12.2.1). Then*

$$f(Z, W) = \sum_{q : (\mathcal{O}/m\mathcal{O})^2} F_q(Z) \vartheta_{m,q}(Z, W),$$

where

$$F_q(Z) = \sum_{T \geq q^t\overline{q}/m} a(T, q) e^{2\pi i(T - q^t\overline{q}/m, Z)}.$$

and

$$\vartheta_{m,q}(Z,W) = \sum_{\substack{h=\lambda+q/m \\ \lambda \in \mathcal{O}^2}} \exp\left\{2\pi im\left[(h^t\overline{h}, Z) + \sigma(h_1, w_2) + \sigma(h_2, w_2)\right]\right\}.$$

Proof. Set $p = q + m\lambda$ with q ranging over all representatives of $(\mathcal{O}/m\mathcal{O})^2$ and λ ranging over $\mathcal{O}^2$ in the first summation of $f(Z, W)$. Then our assertion follows from (J-4) and a direct verification. □

For each $q = {}^t(q_1, q_2) \in \mathcal{O}^2$, consider the *theta series* $\vartheta_{m,q}(Z, W)$ defined by

$$\vartheta_{m,q}(Z,W) = \sum_{\substack{h=\lambda+q/m \\ \lambda \in \mathcal{O}^2}} \exp\left\{2\pi im\left[(h^t\overline{h}, Z) + \sigma(h_1, w_2) + \sigma(h_2, w_2)\right]\right\}.$$

$$(12.2.2)$$

Obviously, one has

$$\vartheta_{m,q}(Z+B, W) = e^{2\pi i(q^t\overline{q}, B)/m}\vartheta_{m,q}(Z, W) \qquad (12.2.3)$$

for all $T \in \Lambda_2$, and

$$\vartheta_{m,q}(Z[U], {}^t\overline{U}W) = \vartheta_{m,Uq}(Z, W) \qquad (12.2.4)$$

for $U = \left[\begin{smallmatrix} 1 & 0 \\ t & 1 \end{smallmatrix}\right]$ or $\left[\begin{smallmatrix} 0 & 1 \\ -1 & 0 \end{smallmatrix}\right]$, $t \in \mathcal{O}$.

Here we prove the transformation formula between $\vartheta_{m,q}(-Z^{-1}, Z^{-1}W)$ and $\vartheta_{m,q}(Z, W)$. We need the following lemmas.

Lemma 12.2.2. *For each* $h = {}^t(h_1, h_2) \in \mathscr{C}_{\mathbb{R}}^2$, $\Lambda = \mathrm{diag}\,[\xi_1, \xi_2]$, $\xi_1 > 0$, $\xi_2 > 0$, *one has*

$$(h^t\overline{h}, \Lambda[U]) = ((Uh)({}^t\overline{h}{}^t\overline{U}), \Lambda)$$

for all $U = \left[\begin{smallmatrix} 1 & t \\ 0 & 1 \end{smallmatrix}\right], \left[\begin{smallmatrix} 1 & 0 \\ t & 1 \end{smallmatrix}\right]$ *or* $\left[\begin{smallmatrix} 0 & 1 \\ -1 & 0 \end{smallmatrix}\right]$, $t \in \mathscr{C}_{\mathbb{R}}$.

Proof. It is obvious for $U = \left[\begin{smallmatrix} 0 & 1 \\ -1 & 0 \end{smallmatrix}\right]$. Here we prove the case $U = \left[\begin{smallmatrix} 1 & t \\ 0 & 1 \end{smallmatrix}\right]$. We have

$$(h^t\overline{h}, \Lambda[U]) = \xi_1 N(h_1) + \xi_2\sigma(t, h_1\overline{h}_2) + (\xi_2 + \xi_1 N(t))N(h_2).$$

On the other hand, we have $Uh = \begin{bmatrix} h_1 + th_2 \\ h_2 \end{bmatrix}$. It follows that

$$((Uh)({}^t\overline{h}{}^t\overline{U}), \Lambda) = \xi_1 N(h_1 + th_2) + \xi_2 N(h_2)$$
$$= \xi_1 N(h_1) + \xi_1\sigma(h_1, th_2) + (\xi_1 N(t_1) + \xi_2)N(h_2).$$

Hence our assertion follows from the fact that

$$\sigma(t, h_1\overline{h}_2) = T(t(h_2\overline{h}_1)) = T((th_2)\overline{h}_1) = \sigma(h_1, th_2). \qquad \square$$

In exact the same way, we prove the following lemma.

Lemma 12.2.3. *For* $h = {}^t(h_1, h_2) \in \mathscr{C}_{\mathbb{R}}^2$, $V = {}^t(v_1, v_2) \in \mathscr{C}_{\mathbb{C}}^2$, $\Lambda = \text{diag}\,[\xi_1, \xi_2]$, $\xi_1 > 0$, $\xi_2 > 0$, *and* $U = \left[\begin{smallmatrix} 1 & t \\ 0 & 1 \end{smallmatrix}\right]$ *or* $\left[\begin{smallmatrix} 1 & 0 \\ t & 1 \end{smallmatrix}\right]$, $t \in \mathscr{C}_{\mathbb{C}}$, *one has*

$$T({}^t\overline{h}(\Lambda^{-1}[{}^t\overline{U}^{-1}]V)) = T(({}^t\overline{h}U^{-1})\Lambda^{-1}({}^t\overline{U}^{-1}V)).$$

Proposition 12.2.4. *Suppose that* $\vartheta_{m,q}(Z, W)$ *is defined as in* (12.2.2). *Then*

$$\vartheta_{m,q}(-Z^{-1}, Z^{-1}W)$$
$$= (\det Z)^4 e^{2\pi i Z^{-1}[W]} \frac{1}{m^8} \sum_{p:(\mathcal{O}/m\mathcal{O})^2} e^{-2\pi i [\sigma(q_1,p_1)+\sigma(q_2,p_2)]/m} \vartheta_{m,p}(Z, W).$$

$$(12.2.5)$$

Proof. It suffices to prove that (12.2.5) holds for $Z = iY$ and $W = iV$ since both sides are holomorphic functions in Z and W. Let $Y = \Lambda[U]$ with $\Lambda = \text{diag}\,[\xi_1, \xi_2]$ and $U = \left[\begin{smallmatrix} 1 & t \\ 0 & 1 \end{smallmatrix}\right]$. Then

$$Y^{-1} = \Lambda^{-1}[{}^tU].$$

It follows that

$$\vartheta_{m,q}(iY^{-1}, Y^{-1}V)$$
$$= \sum_{\substack{h=\lambda+q/m \\ \lambda\in\mathcal{O}^2}} e^{-2\pi m(h^t\overline{h}, Y^{-1})+2\pi i m T(({}^t\overline{h})(Y^{-1}V))}$$
$$= \sum_{\substack{h=\lambda+q/m \\ \lambda\in\mathcal{O}^2}} e^{-2\pi m(h^t\overline{h}, \Lambda^{-1}[{}^tU^{-1}])+2\pi i m T(({}^t\overline{h})(\Lambda^{-1}[{}^tU^{-1}]V))}$$
$$= \sum_{\substack{h=\lambda+q/m \\ \lambda\in\mathcal{O}^2}} e^{-2\pi m(({}^tU^{-1}h)({}^t\overline{h}U^{-1}), \Lambda^{-1})+2\pi i m T(({}^t\overline{h}U^{-1})\Lambda^{-1}({}^t\overline{U}^{-1}V))}$$
$$= e^{-2\pi m\Lambda^{-1}[{}^t\overline{U}^{-1}V]}(\xi_1\xi_2)^4 m^{-8}$$
$$\times \sum_{h\in\mathcal{O}^2} e^{-2\pi m^{-1}((Uh)({}^t\overline{h}{}^t\overline{U}), \Lambda)-2\pi T({}^t\overline{h}{}^t\overline{U})({}^t\overline{U}^{-1})(q/m-iV)}$$

$$= e^{-2\pi m Y^{-1}[V]} (\det Y)^4 m^{-8}$$

$$\times \sum_{h \in \mathcal{O}^2} e^{-2\pi m^{-1}(h^t \overline{h}, Y) - 2\pi i \sigma(q_1/m - iv_1, h_1) - 2\pi i \sigma(q_2/m - iv_2, h_2)}$$

$$= e^{-2\pi m Y^{-1}[V]} (\det Y)^4 m^{-8} \sum_{p:(\mathcal{O}/m\mathcal{O})^2} e^{-2\pi i [\sigma(q_1, p_1) + \sigma(q_2, p_2)]/m}$$

$$\times \sum_{\substack{h = \lambda + p/m \\ \lambda \in \mathcal{O}^2}} e^{-2\pi m (h^t \overline{h}, Y) - 2\pi m (\sigma(v_1, h_1) + \sigma(v_2, h_2))}$$

$$= e^{-2\pi m Y^{-1}[V]} (\det Y)^4 m^{-8}$$

$$\times \sum_{p:(\mathcal{O}/m\mathcal{O})^2} e^{-2\pi i [\sigma(q_1, p_1) + \sigma(q_2, p_2)]/m} \vartheta_{m,p}(iY, iV).$$

This proves our assertion.   □

We then have a result similar to Proposition 10.4.3.

Proposition 12.2.5. *There exists a group homomorphism* $\psi_2 \colon \Gamma_2 \to U(m^{16})$,
where $U(m^{16})$ *is a unitary group of size* m^{16}, *determined by*

(a) $\psi_2(p_B) = \mathrm{diag} \left[e^{-2\pi i (q^t \overline{q}, B)} \right]_{q:(\mathcal{O}/m\mathcal{O})^2}$, $B \in \Lambda_2$;

(b) $\psi_2(t_U) = [s_{p,q}]_{p,q:(\mathcal{O}/m\mathcal{O})^2}$ *with*

$$s_{p,q} = \begin{cases} 1, & \text{if } q = U_p; \\ 0, & \text{otherwise}, \end{cases}$$

for $U = \left[\begin{smallmatrix} 1 & t \\ 0 & 1 \end{smallmatrix}\right]$ *or* $\left[\begin{smallmatrix} 0 & 1 \\ -1 & 0 \end{smallmatrix}\right]$, $t \in \mathcal{O}$;

(c) $\psi_2(\iota) = m^{-8} \left[e^{2\pi i [\sigma(p_1, q_1) + \sigma(p_2, q_2)]/m} \right]_{p,q:(\mathcal{O}/m\mathcal{O})^2}$.

12.3. Jacobi-Eisenstein Series

As shown in Proposition 12.2.1, we are able to decompose a Jacobi form
of degree two into an inner product of a vector-valued modular form and a
vector of theta series. Now with the properties (12.2.3), (12.2.4), (12.2.5)
of theta series defined in (12.2.2), we can characterize a Jacobi form as a
vector-valued modular form.

Proposition 12.3.1. *Let* $\{q_1, q_2, \ldots, q_{m^{16}}\}$ *be a set of representatives of* $(\mathcal{O}/m\mathcal{O})^2$,

$$F(Z) = {}^t(F_{q_1}(Z), F_{q_2}(Z), \ldots, F_{q_{m^{16}}}(Z)) \tag{12.3.1}$$

and

$$\Theta(Z, W) = {}^t(\vartheta_{m,q_1}(Z, W), \vartheta_{m,q_2}(Z, W), \ldots, \vartheta_{m,q_{m^{16}}}(Z, W)) \tag{12.3.2}$$

with $\vartheta_{m,q}(Z, W)$ *as defined in* (12.2.2) *and*

$$F_q(Z) = \sum_{\substack{T \in \Lambda_2 \\ T \geq q^t \bar{q}/m}} \alpha(T, q) e^{2\pi i (T - q^t \bar{q}/m, Z)}.$$

Then the following statements are equivalent.

(1) $f(Z, W) = {}^t F(Z) \Theta(Z, W)$ *is a Jacobi form of weight* k *and index* m *with respect to* Γ_2.

(2) $F(Z)$ *satisfies the following conditions:*

   (a) $F(Z + B) = \psi_2(p_B) F(Z)$ *for* $B \in \Lambda_2$;

   (b) $F(Z[U]) = \psi_2(t_U) F(Z)$ *for* $U = \begin{bmatrix} 1 & t \\ 0 & 1 \end{bmatrix}$ *or* $\begin{bmatrix} 0 & 1 \\ -1 & 0 \end{bmatrix}$, $t \in \mathcal{O}$;

   (c) $F(-Z^{-1}) = (\det Z)^{k-4} \psi_2(\iota) F(Z)$.

Proof. It is similar to the proof of Proposition 10.5.1, so we omit it here. $\square$

Now we use the group homomorphism ψ_2 to construct a vector-valued modular form corresponding to a Jacobi form of degree two. Let $j(g, Z)$ be a factor of the determinant of Jacobi matrix of $g \in \Gamma_2$ at $Z \in \mathcal{H}_2$ determined by the following.

(1) $j(p_B, Z) = 1$ for all $B \in \Lambda_2$;

(2) $j(t_U, Z) = 1$ for all $U = \begin{bmatrix} 1 & t \\ 0 & 1 \end{bmatrix}$ or $\begin{bmatrix} 0 & 1 \\ -1 & 1 \end{bmatrix}$, $t \in \mathcal{O}$;

(3) $j(\iota, Z) = \det(-Z)$;

(4) $j(g_1 g_2, Z) = j(g_1, g_2(Z)) j(g_2, Z)$.

Also let Γ_2^0 be the subgroup of Γ_2 generated by p_B and t_U, $B \in \Lambda_2$, $U = \left[\begin{smallmatrix} 1 & t \\ 0 & 1 \end{smallmatrix}\right]$ or $\left[\begin{smallmatrix} 0 & 1 \\ -1 & 0 \end{smallmatrix}\right]$, $t \in \mathcal{O}$. For each $q \in \mathcal{O}^2$ with $q^t\bar{q} \equiv 0 \pmod{m}$, we define

$$E_{k,m}(Z,W;q) = {}^t(E_{q,q_1}(Z), E_{q,q_2}(Z), \ldots, E_{q,q_{m^{16}}}(Z))\Theta(Z,W) \quad (12.3.3)$$

with

$$E_{q,p}(Z) = \sum_{M:\Gamma_2/\Gamma_2^0} j(M,Z)^{4-k}\overline{\psi_{q,p}(M)}, \quad (12.3.4)$$

where

$$\psi(M) = (\psi_{q,p}(M))_{q,p:(\mathcal{O}/m\mathcal{O})^2}. \quad (12.3.5)$$

The series (12.3.4) converges absolutely and uniformly on compact subset of $\mathcal{H}^2$ if $k > 22$. Here we shall prove that the vector-valued modular form corresponding to $E_{k,m}(Z,W;q)$ satisfies condition (2) of Proposition 12.3.1. Consequently, $E_{k,m}(Z,W;q)$ is indeed a Jacobi form of weight k and index m for $k > 22$ and $q^t\bar{q} \equiv 0 \pmod{m}$.

Proposition 12.3.2. *For $k > 22$ and $q \in \mathcal{O}^2$ with $q^t\bar{q} \equiv 0 \pmod{m}$, the Jacobi-Eisenstein defined in (12.3.3) is a Jacobi form of weight k and index m.*

Proof. Let $E(Z;q)$ be the vector-valued modular form corresponding to $E_{k,m}(Z,W;q)$. Then ${}^tE(Z;q)$ is the qth row of the matrix

$$\sum_{M:\Gamma_2/\Gamma_2^0} j(M,Z)^{4-k}\overline{\psi_2(M)}.$$

Thus the condition (b) follows from the cocycle condition of $j(M,Z)$ and the properties of ψ_2.   □

12.4. Applications to Singular Modular Forms on the Exception Domain

Besides Jacobi-Eisenstein series as constructed in previous section, the Jacobi-Fourier coefficients of modular forms on the exceptional domain provide another kind of examples for Jacobi forms of degree two. Here we shall determine explicitly the Fourier coefficients of modular forms of weight 4 and 8 on the exceptional domain.

Proposition 12.4.1. *Let $E_4(Z)$ be a modular form of weight 4 on the exceptional domain with the Fourier expansion*

$$E_4(Z) = \sum_{T \in \mathcal{T}_O \cap \mathcal{K}^+} a(T) e^{2\pi i (T,Z)}. \qquad (12.4.1)$$

Then $a(T) = 0$ unless $\operatorname{rank} T \leq 1$. If $a(0) = 1$, then for $\operatorname{rank} T = 1$,

$$a(T) = 240 \sum_{d \mid \epsilon(T)} d^3, \qquad (12.4.2)$$

where $\epsilon(T)$ is the largest integer d such that $d^{-1}T \in \mathcal{T}_O$.

Proof. Let

$$\psi_0(Z_1) + \sum_{m=1}^{\infty} \varphi_m(Z_1, W) e^{2\pi i m z_3}$$

be the Jacobi-Fourier expansion of $E_4(Z)$ with

$$Z = \begin{bmatrix} Z_1 & W \\ {}^t\overline{W} & z_3 \end{bmatrix}.$$

Then $\varphi_0(Z_1)$ is a modular form of weight 4 on $\mathcal{H}_2$, hence it is a constant multiple of $f_4(Z_1)$ in Proposition 10.1.1. Note that $a(0) = 1$, it follows that $\varphi_0(Z_1) = f_4(Z_1)$ and $a \left(\begin{smallmatrix} T_1 & 0 \\ 0 & 0 \end{smallmatrix} \right)$ is given by $240 \sum_{d \mid \epsilon(T_1)} d^3$ if $\det T_1 = 0$, $T_1 \neq 0$.

On the other hand, $\varphi_m(Z_1, W)$ is a Jacobi form of weight 4 and index m on $\mathcal{H}_2 \times \mathscr{C}_{\mathbb{C}}^2$. By Proposition 12.3.1, we decompose $\varphi_m(Z_1, W)$ into

$$\sum_{q:(\mathcal{O}/m\mathcal{O})^2} F_q(Z_1) \vartheta_{m,q}(Z_1, W)$$

with

$$F_q(Z_1) = \sum_{T \in \Lambda_2 T \geq q^t \overline{q}/m} a \begin{pmatrix} T_1 & q \\ {}^t\overline{q} & m \end{pmatrix} e^{2\pi i (T - q^t \overline{q}/m, Z_1)}.$$

By Proposition 12.3.1, we know that

$$F(Z_1) = {}^t F_q(Z_1)_{q:(\mathcal{O}/m\mathcal{O})^2}$$

is a vector-valued modular form of weight 0. It forces that $F_q(Z_1)$ is a constant and hence

$$a \begin{pmatrix} T_1 & q \\ {}^t\overline{q} & m \end{pmatrix} = 0$$

unless $T_1 = q^t \bar{q}/m$. This proves that $a(T) = 0$ unless $\operatorname{rank} T \leq 1$. For $\mathscr{T}_{\mathcal{O}}$ with $\operatorname{rank} T = 1$, we are able to reduce T to $T_0 = \operatorname{diag}[\epsilon(T), 0, 0]$ by a finite number of operations $T \mapsto T[U]$ with $U = U_{ij}(t)$, $i \neq j$ and $t \in \mathcal{O}$. Thus

$$a(T) = a(T_0) = 240 \sum_{d \mid \epsilon(T)} d^3. \qquad \qquad \square$$

Next, we give a relation among Fourier coefficients of the modular form of weight 4.

Proposition 12.4.2. *For positive integer m and $q = {}^t(q_1, q_2) \in \mathcal{O}^2$, let*

$$T = \begin{bmatrix} q^t \bar{q}/m & q \\ {}^t\bar{q} & m \end{bmatrix}$$

and

$$G_m(q) = \begin{cases} 240 \sum_{d \mid \epsilon(T)} d^3, & \text{if } T \in \mathscr{T}_{\mathcal{O}}; \\ 0, & \text{otherwise.} \end{cases}$$

Then

$$G_m(q) = \frac{1}{m^8} \sum_{p:(\mathcal{O}/m\mathcal{O})^2} e^{-2\pi i[(q_1, p_1) + (q_2, p_2)]/m} G_m(p). \qquad (12.4.3)$$

Proof. It follows from the fact that $G = {}^t(G_m(q))_{q:(\mathcal{O}/m\mathcal{O})^2}$ is the vector-valued modular form corresponding to $\varphi_m(Z_1, W)$, the mth Jacobi-Fourier coefficient of $E_4(Z)$. Thus G must satisfy the condition (b) of Proposition 12.3.1. In particular,

$$G = \psi_2(\iota)G. \qquad (12.4.4)$$

This is precisely the identity (12.4.3) in vector form. $\qquad \square$

Remark 12.4.3. The identity (12.4.3) was proved in [12] directly from the definition of $G_m(q)$ and it implies that

$$\varphi_m(Z, W) = \sum_{q:(\mathcal{O}/m\mathcal{O})^2} G_m(q)\vartheta_{m,q}(Z_1, W), \quad (Z_1, W) \in \mathcal{H}_2 \times \mathscr{C}_{\mathbb{C}}^2$$

is a Jacobi form of weight 4 and index m. With $\varphi_m(Z_1, W)$ as the m-th coefficient, we are able to define a holomorphic function on the exceptional domain as

$$E(Z) = f_4(Z) + \sum_{m=1}^{\infty} \varphi_m(Z_1, W)e^{2\pi i m z_3}, \quad Z = \begin{bmatrix} Z_1 & W \\ {}^t\overline{W} & z_3 \end{bmatrix} \in \mathcal{H}.$$

With the theory of Jacobi forms on $\mathcal{H}_1 \times \mathscr{C}_\mathbb{C}$ as well as Jacobi forms on $\mathcal{H}_2 \times \mathscr{C}_\mathbb{C}^2$, it is easy to verify that

$$E(-Z^{-1}) = (\det Z)^4 E(Z). \tag{12.4.5}$$

Consequently, we provide another way to construct the singular modular form of weight 4 on the exceptional domain.

Note that $E^2(Z)$ is a modular form of weight 8. Indeed it is a singular modular form of weight 8 and its Fourier coefficients can be determined explicitly by the following proposition.

Proposition 12.4.4. *Let*

$$E^2(Z) = \sum_{T \in \mathscr{T}_0 \cap \mathcal{K}^+} b(T) e^{2\pi i (T, Z)}.$$

Then $b(T) = 0$ unless $\operatorname{rank} T \leq 2$ *and*

$$b\left(\begin{bmatrix} T_1 & 0 \\ 0 & 0 \end{bmatrix}\right) = \begin{cases} 1, & \text{if } T_1 = 0; \\ 480 \sum_{d | \epsilon(T_1)} d^7, & \text{if } \det T_1 = 0, \ T_1 \neq 0; \\ 240 \cdot 480 \sum_{d | \epsilon(T_1)} d^7 \sum_{d_1 | \det(d^{-1} T_1)} d_1^3, & \text{if } \det T_1 \neq 0. \end{cases}$$

Proof. The Fourier coefficient $a(T)$ of $E(Z)$ has the property that $a(T) = 0$ unless $\operatorname{rank} T \leq 1$. It follows that

$$b(T) = \sum_{T_1 + T_2 = T} a(T_1) a(T_2)$$

is zero unless $\operatorname{rank} T \leq 2$. Let

$$\psi_0(Z_1) + \sum_{m=1}^{\infty} \psi_m(Z_1, W) e^{2\pi i m z_3}$$

be the Jacobi-Fourier expansion of $E^2(Z)$. Then

$$\psi_0(Z_1) = \sum_{T_1 \in \Lambda_2} b\left(\begin{bmatrix} T_1 & 0 \\ 0 & 0 \end{bmatrix}\right) e^{2\pi i (T_1, Z_1)} = \lim_{\lambda \to \infty} E^2\left(\begin{bmatrix} Z_1 & 0 \\ 0 & i\lambda \end{bmatrix}\right)$$

is a modular form of weight 8 and hence it is equal to $[f_4(Z_1)]^2$. But $[f_4(Z_1)]^2$ is the only modular form of weight 8 which is also in the Maaß space, its coefficients satisfy the Maaß condition. So it suffices to know $b\left(\left[\begin{smallmatrix} T_1 & 0 \\ 0 & 0 \end{smallmatrix}\right]\right)$ with $T_1 = \left[\begin{smallmatrix} n & 0 \\ 0 & 0 \end{smallmatrix}\right]$ or $T_1 = \left[\begin{smallmatrix} n & t \\ \bar{t} & 1 \end{smallmatrix}\right]$, $n - N(t) \neq 0$. Note that

$$b\left(\begin{bmatrix} T_1 & 0 \\ 0 & 0 \end{bmatrix}\right) = \#\left\{h_1, h_2 \in \mathcal{O}^2 \mid h_1{}^t\bar{h}_1 + h_2{}^t\bar{h}_2 = T_1\right\}.$$

For $T_1 = \left[\begin{smallmatrix} n & 0 \\ 0 & 0 \end{smallmatrix}\right]$, we have

$$b\left(\begin{bmatrix} T_1 & 0 \\ 0 & 0 \end{bmatrix}\right) = \#\left\{a, b \in \mathcal{O} \mid N(a) + N(b) = n\right\}$$
$$= 480 \sum_{d|n} d^7.$$

On the other hand, for $T_1 = \left[\begin{smallmatrix} n & t \\ \bar{t} & 1 \end{smallmatrix}\right]$, we have

$$b\left(\begin{bmatrix} T_1 & 0 \\ 0 & 0 \end{bmatrix}\right) = \#\left\{a, b, c \in \mathcal{O} \mid N(c) = 1, N(a) + N(b) = n, a\bar{c} = t\right\}$$
$$= 240 \cdot 480 \sum_{d|(n-N(t))} d^3.$$

$\square$

12.5. Jacobi Cusp Forms

A Jacobi form f of weight k and index m with the Fourier expansion

$$f(Z, W) = \sum_{T \geq q^t\bar{q}} \sum_{q \in (\mathcal{O}/m\mathcal{O})^2} a(T, q) e^{2\pi i[(T, Z) + \sigma(q_1, w_1) + \sigma(q_2, w_2)]}$$

is called a *Jacobi cusp form* if it satisfies the further condition

$$a(T, q) = 0 \quad \text{if} \quad \det\left(T - \frac{1}{m} q^t\bar{q}\right) = 0. \tag{12.5.1}$$

Let ${}^t(F_q(Z))_{q:(\mathcal{O}/m\mathcal{O})^2}$ and ${}^t(G_q(Z))_{q:(\mathcal{O}/m\mathcal{O})^2}$ be the modular forms corresponding to Jacobi forms f and g of weight k and index m, respectively. For $k > 22$, when at least one of f and g is a Jacobi form, we define the inner product of f and g by

$$\langle f, g \rangle = \int_{\Gamma_2 \backslash \mathcal{H}_2} (\det Y)^{k-14} \sum_{q:(\mathcal{O}/m\mathcal{O})^2} F_q(Z)\overline{G_q(Z)} \, dX \, dY, \tag{12.5.2}$$

where $\Gamma_2 \setminus \mathcal{H}_2$ is a fundamental domain of $\mathcal{H}_2$ under the action of Γ_2 and $dXdY$ is the usual Euclidean measure on $\mathcal{H}_2$, i.e.,

$$dXdY = dx_1 dx_2 dx_{12} dy_1 dy_2 dy_{12}, \quad Z = \begin{bmatrix} x_1 & x_{12} \\ \overline{x}_{12} & x_2 \end{bmatrix} + i \begin{bmatrix} y_1 & y_{12} \\ \overline{y}_{12} & y_2 \end{bmatrix}.$$

Note that $(\det Y)^{-10} \, dXdY$ is an invariant measure on $\mathcal{H}_2$.

With the inner product as given in (12.5.2), a Jacobi cusp form is orthogonal to a Jacobi-Eisenstein series of the same weight and index.

Proposition 12.5.1. *Let f be a Jacobi cusp form of weight k and index m and $E_{k,m}(Z, W; q)$ be the Jacobi-Eisenstein series defined in (12.3.3) for $q^t \overline{q} \equiv 0 \pmod{m}$. Then for $k > 22$, we have*

$$\langle E_{k,m}, f \rangle = 0.$$

Proof. Let

$$^t(F_q(Z))_{q:(\mathcal{O}/m\mathcal{O})^2}$$

be the vector-valued modular form of f and $\alpha(T, q)$ be the Fourier coefficient of f. Then we have

$$F_q(Z) = \sum_{T \geq m^{-1} q^t \overline{q}} \alpha(T, q) e^{2\pi i (T - q^t \overline{q}/m, Z)}.$$

For $q^t \overline{q} \equiv 0 \pmod{m}$, we have $q^t \overline{q}/m \in \Lambda_2$ and for all $T > q^t \overline{q}/m$,

$$\int_{\mathscr{T}_{\mathbb{R}}^{(2)} / \Lambda_2} e^{-2\pi i (T - q^t \overline{q}/, X)} \, dX = 0,$$

where $\mathscr{T}_{\mathbb{R}}^{(2)}$ is the set of 2×2 Hermitian matrices over real Cayley numbers. Thus it follows that

$$\int_{\Gamma_2^0 \setminus \mathcal{H}_2} (\det Y)^{k-14} \overline{F_q(Z)} \, dXdY = 0$$

since we are able to construct a fundamental domain with its real part given by $\mathscr{T}_{\mathbb{R}}^{(2)} / \Lambda_2$ on $\mathcal{H}_2$ for Γ_2^0. Hence

$$\int_{\Gamma_2 \setminus \mathcal{H}_2} (\det Y)^{k-14} \sum_{M : \Gamma_2 / \Gamma_2^0} |j(M, Z)|^{8-2k} \overline{F_q(M(Z))} \, dXdY = 0.$$

In light of the formula

$$F_q(M(Z)) = \sum_{p:(\mathcal{O}/m\mathcal{O})^2} j(M, Z)^{k-4} \psi_{2,p,q}(M) F_q(Z),$$

we get our assertion.   □

Next, we shall count the number of different Jacobi-Eisenstein series defined in (12.3.3). For $m = p^\nu$, by the proof of Proposition 5 of [12], we have

$$\#\left\{q \in (\mathcal{O}/m\mathcal{O})^2 \mid q^t\bar{q} \equiv 0 \quad (\mathrm{mod}\ m)\right\} = p^{8\nu}\left\{\sum_{\tau=0}^{\nu} p^{3\tau} - \sum_{\tau=0}^{\nu-1} p^{3\tau-5}\right\}.$$
$$(12.5.3)$$

Set

$$N_m = \#\left\{q \in (\mathcal{O}/m\mathcal{O})^2 \mid q^t\bar{q} \equiv 0 \quad (\mathrm{mod}\ m), 2q \equiv 0 \quad (\mathrm{mod}\ m)\right\}.$$
$$(12.5.4)$$

By (12.5.3) and an elementary consideration yields

$$N_m = \begin{cases} 1, & \text{if } m \equiv 1 \pmod 2; \\ 2240, & \text{if } m \equiv 2 \pmod 4; \\ 65536, & \text{if } m \equiv 0 \pmod 4. \end{cases} \qquad (12.5.5)$$

Now fix a particular set of representatives of $(\mathcal{O}/m\mathcal{O})^2$ as follows:

$$q_1, \ldots, q_r, q_{r+1}, \ldots, q_{r+s}, -q_{r+1}, \ldots, -q_{r+s}, \quad r = N_m, \quad r + 2s = m^{16},$$

with $2q_j \equiv 0 \pmod m$ for $1 \le j \le r$, and $2q_\ell \not\equiv 0 \pmod m$ for $r + 1 \le \ell \le r + s$. With the above as index set of $(\mathcal{O}/m\mathcal{O})^2$, we have the following proposition.

Proposition 12.5.2. *Let ι be the transform $Z \mapsto -Z^{-1}$. Then*

$$\psi_2(\iota^2) = \begin{bmatrix} E_r & 0 & 0 \\ 0 & 0 & E_s \\ 0 & E_s & 0 \end{bmatrix},$$

where E_ℓ is the identity matrix of size $\ell \times \ell$.

Proof. Let $s_{p,q}$ be the entry of $\psi_2(\iota^2)$ at (p,q) position. Then

$$
\begin{aligned}
s_{p,q} &= \frac{1}{m^{16}} \sum_{r:(\mathcal{O}/m\mathcal{O})^2} e^{2\pi i[\sigma(p_1,r_1)+\sigma(p_2,r_2)]/m} e^{2\pi i[\sigma(r_1,q_1)+\sigma(r_2,q_2)]/m} \\
&= \frac{1}{m^{16}} \sum_{r_1:\mathcal{O}/m\mathcal{O}} e^{2\pi i\sigma(p_1+q_1,r_1)/m} \sum_{r_2:\mathcal{O}/m\mathcal{O}} e^{2\pi i\sigma(p_2+q_2,r_2)/m} \\
&= \begin{cases} 1, & \text{if } p_1 + q_1 \equiv 0 \pmod{m} \text{ and } p_2 + q_2 \equiv 0 \pmod{m}; \\ 0, & \text{otherwise.} \end{cases}
\end{aligned}
$$

$\square$

Proposition 12.5.3. *For any $M \in \Gamma_2$, $\psi_2(M)$ appears to be the form*

$$
\begin{bmatrix} A_r & H_{rs} & H_{rs} \\ F_{sr} & B_s & G_s \\ F_{sr} & G_s & B_s \end{bmatrix},
$$

where A_r is a matrix of size $r \times r$, H_{rs} is a matrix of size $r \times s$, and F_{sr} is a matrix of size $s \times r$. In particular, one has

$$
\psi_{2,-q,p}(M) = \psi_{2,p,-p}(M).
$$

Proof. It follows from $\psi_2(M)$ commutes with $\psi_2(\iota^2) = \psi_2(t_U)$ with $U = -E_2$. $\square$

Proposition 12.5.4. *For any $q \in \mathcal{O}^2$ with $q^t\bar{q} \equiv 0 \pmod{m}$, one has*

$$
E_{k,m}(Z,W;-q) = E_{k,m}(Z,W;q).
$$

Proof.

$$
\begin{aligned}
E_{k,m}(Z,W;-q) &= \sum_{M:\Gamma_2/\Gamma_2^0} j(M,Z)^{4-k} \sum_{p:(\mathcal{O}/m\mathcal{O})^2} \overline{\psi_{2,-q,p}(M)} \vartheta_{m,p}(Z,W) \\
&= \sum_{M:\Gamma_2/\Gamma_2^0} j(M,Z)^{4-k} \sum_{p:(\mathcal{O}/m\mathcal{O})^2} \overline{\psi_{2,q,-p}(M)} \vartheta_{m,p}(Z,W) \\
&= \sum_{M:\Gamma_2/\Gamma_2^0} j(M,Z)^{4-k} \sum_{p:(\mathcal{O}/m\mathcal{O})^2} \overline{\psi_{2,q,p}(M)} \vartheta_{m,-p}(Z,W) \\
&= \sum_{M:\Gamma_2/\Gamma_2^0} j(M,Z)^{4-k} \sum_{p:(\mathcal{O}/m\mathcal{O})^2} \overline{\psi_{2,q,p}(M)} \vartheta_{m,p}(Z,-W)
\end{aligned}
$$

$$= E_{k,m}(Z, -W; q)$$
$$= E_{k,m}(Z[-E_2], (-E_2)W; q)$$
$$= E_{k,m}(Z, W; q).$$

$\square$

By (12.5.3), (12.5.5) and the previous proposition, we have the following proposition.

Proposition 12.5.5. *The total number of different Jacobi-Eisenstein series is give by*

$$\frac{1}{2}\left\{ m^8 \prod_{p|m} \left(\sum_{\tau=0}^{\nu_p(m)} p^{3\tau} - \sum_{\tau=0}^{\nu_p(m)-1} p^{3\tau-5} \right) + N_m \right\},$$

where $\nu_p(m)$ is the p-adic valuation defined by $\nu_p(m) = \alpha$ if p^α is the highest power of p dividing m.

The set of Jacobi-Eisenstein series is the orthogonal complement of the vector space of Jacobi cusp form which is denoted by $J_{k,m}^0(\Gamma_2)$ with respect to the inner product (12.5.2). By realizing Jacobi cusp forms as vector-valued cusp forms, we are able to compute its dimension via Selberg trace formula as we had done in [12]. For $k > 22$,

$$\dim J_{k,m}^0(\Gamma_2)$$

$$= c(k) \int_{\Gamma_2 \backslash \mathcal{H}_2} (\det Y)^{k-14} \sum_{M \in \Gamma_2} \det \left[\frac{1}{2i} \left(Z - \overline{M(Z)} \right) \right]^{4-k} \qquad (12.5.6)$$

$$\times \overline{j(M, Z)}^{4-k} \operatorname{tr}(\psi_2(M)) \, dX dY,$$

where

$$c(k) = 2^{-13} \pi^{-10} \frac{\Gamma(k-4)\Gamma(k-8)}{\Gamma(k-9)\Gamma(k-13)}.$$

The leading term in (12.5.6) is the total contribution from positive and minus identity of Γ_2, which is given by

$$I(\pm E) = c(k) \int_{\Gamma_2 \backslash \mathcal{H}_2} (\det Y)^{-10} \, dX dY \cdot \left\{ m^{16} + \gcd(m, z)^{16} \right\}. \qquad (12.5.7)$$

Of course, the formula (12.5.6) is still far away an explicit dimension formula for the vector space $J^0_{k,m}(\Gamma_2)$ since we only know the contribution from positive and minus identity. However, we can use it to compute explicitly the contribution from a particular conjugacy class of Γ_2 and hence obtain an approximate formula for $\dim J^0_{k,m}(\Gamma_2)$.

Appendix A

The p-Adic Numbers

A.1. The p-Adic Integers

For any positive integer n and prime number p, let $A_n = \mathbb{Z}/p^n\mathbb{Z}$ be the ring of residue classes modulo p^n. Suppose $\varphi_n \colon A_n \to A_{n-1}$ is the natural projection. Then φ_n is an onto mapping and its kernel is $p^{n-1}A_n$.

The p-adic integers $\mathbb{Z}_p$ is a subring of $\prod_{n=1}^{\infty} A_n$ and it consists of sequence

$$x = (x_1, x_2, \ldots, x_n, \ldots) \in \prod_{n=1}^{\infty} A_n$$

with $\varphi_n(x_n) = x_{n-1}$, i.e., $x_n \equiv x_{n-1} \pmod{p^{n-1}}$. The ring $\mathbb{Z}_p$ is also called the *inverse limit* of (A_n, φ_n).

Proposition A.1.1. *Let $\epsilon_n \colon \mathbb{Z}_p \to A_n$ be the projection of a p-adic integer into its nth component and $p^n \colon \mathbb{Z}_p \to \mathbb{Z}_p$ be the multiplication by p^n. Then the sequence*

$$0 \longrightarrow \mathbb{Z}_p \xrightarrow{p^n} \mathbb{Z}_p \xrightarrow{\epsilon_n} A_n \longrightarrow 0$$

is an exact sequence of abelian groups.

Proof. First we prove that the multiplication by p from $\mathbb{Z}_p$ into $\mathbb{Z}_p$ is one-to-one. Then it will follow that the multiplication by p^n is also one-to-one. If $x = (x_1, x_2, \ldots, x_n, \ldots) \in \mathbb{Z}_p$ and $px = 0$, then $px_{n+1} \equiv 0 \pmod{p^{n+1}}$.

219

So $x_{n+1} = p^n y_{n+1}$ with $y_{n+1} \in A_{n+1}$. Hence

$$x_n = \varphi_{n+1}(p^n y_{n+1}) \equiv 0 \pmod{p^n}.$$

It is clear that $p^n \mathbb{Z}_p \subseteq \ker(\epsilon_n)$. Conversely, if $x = (x_1, x_2, \ldots, x_n, \ldots) \in \ker(\epsilon_n)$, then $x_m \equiv 0 \pmod{p^n}$ for all $m \geq n$. So there exists a $y_{m-n} \in A_{m-n}$ such that $x_m = p^n y_{m-n}$ for $m \geq n+1$. Let $y = (y_1, y_2, \ldots, y_n, \ldots)$. Then we want to prove $p^n y = x$. It suffices to prove that $p^n y_k \equiv x_k \pmod{p^k}$. It is true for $0 \leq k \leq n$. Now suppose $m \geq n+1$. Then

$$x_m \equiv p^n y_{m-n} \pmod{p^m}.$$

But on the other hand,

$$x_m \equiv p^n y_{m-n} \equiv p^n y_m \pmod{p^m}$$

since $y_{m-n} \equiv y_m \pmod{p^{m-n}}$. This proves our assertion. $\qquad\square$

Remark A.1.2. By the above exact sequence, we can identity $\mathbb{Z}_p/p^n\mathbb{Z}_p$ with $A_n = \mathbb{Z}/p^n\mathbb{Z}$.

Note that $\mathbb{Z}_p$ is a commutative ring with identity $1 = (1, 1, \ldots, 1, \ldots)$ under the componentwise addition and multiplication, i.e.,

$$(x_1, \ldots, x_n, \ldots) + (y_1, \ldots, y_n, \ldots) = (x_1 + y_1, \ldots, x_n + y_n, \ldots),$$
$$(x_1, \ldots, x_n, \ldots)(y_1, \ldots, y_n, \ldots) = (x_1 y_1, \ldots, x_n y_n, \ldots).$$

We also consider $\mathbb{Z}$ as a subring of $\mathbb{Z}_p$ via the embedding

$$x \longrightarrow (x, \ldots, x, \ldots), \quad x \in \mathbb{Z}.$$

Proposition A.1.3. *For any $x = (x_1, x_2, \ldots, x_n, \ldots)$ in $\mathbb{Z}_p$, x is invertible if and only if x_1 is not divisible by p.*

Proof. Suppose $x = (x_1, x_2, \ldots, x_n, \ldots) \in \mathbb{Z}_p$ and it is not divisible by p. Then x_1 is nonzero in $A_1 = \mathbb{Z}/p\mathbb{Z}$. So there exists a y_1 in A_1 such that

$$x_1 y_1 = 1 - pz \quad \text{and} \quad x_n y_1 = 1 - pu.$$

It follows that

$$x_n y_1 (1 + pu + \cdots + p^{n-1}u^{n-1}) = 1 - p^n u^n \equiv 1 \pmod{p^n}.$$

Now let $y_n = y_1(1 + pu + \cdots + p^{n-1}u^{n-1})$ and $y = (y_1, y_2, \ldots, y_n, \ldots)$, then y is the inverse of x in $\mathbb{Z}_p$ by a direct verification. $\qquad\square$

Corollary A.1.4. *The element $x = (x_1, x_2, \ldots, x_n, \ldots) \in \mathbb{Z}_p$ is invertible if $x_1 \neq 0$ in $\mathbb{Z}/p\mathbb{Z}$.*

Proposition A.1.5. *Every p-adic integer x, distinct from zero, has a unique representation of the form*

$$x = p^m u,$$

where m is a nonnegative integer and u is a unit of $\mathbb{Z}_p$.

Proof. Let $x = (x_1, x_2, \ldots, x_n, \ldots)$. If x is a unit, then $m = 0$. Suppose x is not a unit, the previous proposition implies $x_1 \equiv 0 \pmod{p}$. Since x is not zero, the congruence $x_n \equiv 0 \pmod{p^n}$ does not hold for all n. Let m be the largest positive integer such that

$$x_j \equiv 0 \pmod{p^j}, \quad 1 \leq j \leq m.$$

For any positive integer n, one has

$$x_{m+n} \equiv x_m \equiv 0 \pmod{p^m}.$$

Therefore $u_n = x_{m+n}/p^m$ is an integer. From the congruence

$$p^m u_{n+1} - p^m u_n = x_{m+n+1} - x_{m+n} \equiv 0 \pmod{p^{m+n}},$$

we get $u_{n+1} - u_n \equiv 0 \pmod{p^n}$. Thus $u = (u_1, u_2, \ldots, u_n, \ldots)$ is a unit of $\mathbb{Z}_p$ since $u_1 \not\equiv 0 \pmod{p}$. It follows that $x = p^m u$ has the desired representation.

Suppose that x has another representation $p^k \eta$. If $\eta = (\eta_1, \eta_2, \ldots, \eta_n, \ldots)$, then

$$p^m u_n \equiv p^k \eta_n \pmod{p^n}. \tag{A.1.1}$$

But p never divides u_n or η_n. Setting $n = m$ in (A.1.1), we obtain

$$p^m u_m \equiv p^k \eta_m \equiv 0 \pmod{p^m},$$

from which we conclude that $k \geq m$. By symmetry, we also have $m \geq k$, and hence $k = m$. In (A.1.1), replacing n by $n + m$ and dividing by p^m, we obtain

$$u_{m+n} \equiv \eta_{m+n} \pmod{p^n}.$$

Also $u_{m+n} \equiv u_n \pmod{p^n}$ and $\eta_{m+n} \equiv \eta_n \pmod{p^n}$. It follows that $u_n \equiv \eta_n \pmod{p^n}$. Thus $u = \eta$ and the proposition is proved. $\qquad\square$

Corollary A.1.6. *The p-adic integer $x = (x_1, x_2, \ldots, x_n, \ldots)$ is divisible by p^m if and only if $x_n \equiv 0 \pmod{p^n}$ for all $1 \leq n \leq m$ and $x_{m+1} \not\equiv 0 \pmod{p^{m+1}}$.*

Corollary A.1.7. *The ring $\mathbb{Z}_p$ is an integral domain, it has no zero divisors.*

Let U be the group of invertible elements in $\mathbb{Z}_p$, called the group of p-adic units. Every nonzero element x in $\mathbb{Z}_p$ can be written as $p^m u$, $u \in U$, $m \geq 0$.

A.2. The p-Adic Valuation

Let F be a field. A function v defined on F is called a *valuation* of F if it satisfies the following conditions:

(i) $v(\alpha)$ takes all integral values as α ranges over all nonzero elements of F and $v(0) = +\infty$;

(ii) $v(\alpha\beta) = v(\alpha) + v(\beta)$;

(iii) $v(\alpha + \beta) \geq \min(v(\alpha), v(\beta))$.

Let p be a prime and a be a nonzero integer. We define $v_p(a) = n$ if $a = p^n u$ with $(u, p) = 1$. For any rational number a/b with $a, b \in \mathbb{Z}$ and $b \neq 0$, we define

$$v_p\left(\frac{a}{b}\right) = v_p(a) - v_p(b).$$

Also, we set $v_p(0) = \infty$. Then v_p is a valuation of $\mathbb{Q}$. It is called the p-adic valuation.

Exercise A.2.1. Prove that

$$v_p(p^n!) = 1 + p + p^2 + \cdots + p^{n-1}.$$

Exercise A.2.2. Prove that if $n = a_0 + a_1 p + \cdots + a_k p^k$ with $0 \leq a_i \leq p-1$, and if $S_n = a_0 + a_1 + \cdots + a_k$, then

$$v_p(n!) = \frac{n - S_n}{p - 1}.$$

For each prime number p, we define $|\cdot|_p$ on $\mathbb{Q}$ as follows:

$$|x|_p = \begin{cases} p^{-v_p(x)}, & \text{if } x \neq 0; \\ 0, & \text{if } x = 0. \end{cases}$$

It is a direct verification to show that $|\cdot|_p$ is a norm on $\mathbb{Q}$, i.e., it satisfies the following conditions:

(a) $|x|_p = 0$ if and only if $x = 0$;

(b) $|xy|_p = |x|_p |y|_p$;

(c) (Triangle Inequality) $|x + y|_p \leq |x|_p + |y|_p$.

Furthermore, it satisfies

$$|x + y|_p \leq \max\left\{|x|_p, |y|_p\right\}.$$

Such a norm is called non-Archimedean. Two norms $\|\cdot\|_1$ and $\|\cdot\|_2$ on the same space X are called equivalent if there exist two positive constants c_1 and c_2 such that

$$c_1 \|x\|_2 \leq \|x\|_1 \leq c_2 \|x\|_2 \quad \text{for all } x \in X.$$

In the definition of $|\cdot|_p$, if we replace $p^{-v_p(x)}$ by $\rho^{v_p(x)}$ with $0 < \rho < 1$, then we obtain an equivalent non-Archimedean norm. Besides non-Archimedean norm $|\cdot|_p$ on $\mathbb{Q}$, we also have the usual absolute value $|\cdot|$. Sometimes we denote such absolute value by $|\cdot|_\infty$. Hence for any $x \in \mathbb{Q}$, $x \neq 0$,

$$|x|_\infty \prod_p |x|_p = 1.$$

In the following, we prove that there is no other norm on $\mathbb{Q}$.

Theorem A.2.3 (Ostrowski). *Every nontrivial norm $\|\cdot\|$ on $\mathbb{Q}$ is equivalent to $|\cdot|_p$ or $|\cdot|_\infty$.*

Proof. According to the values of positive integers, we divide it into two cases as follows:

Case I. Suppose that $\|n\| \leq 1$ for all positive integers n. Let n_0 be the least positive integer such that $\|n_0\| < 1$. We claim that n_0 must be a prime. Assume $n_0 = ab$ with a, $b < n_0$. So $\|a\| = 1$, $\|b\| = 1$ and $\|n_0\| = \|a\| \|b\| = 1$, a contradiction. Let $n_0 = p$. Next we claim that $\|q\| = 1$ if q is a prime different from p. Suppose not, then $\|q\| < 1$ and

$$\left\|q^N\right\| = \|q\|^N < \frac{1}{2}$$

for some positive integer N. On the other hand, we also have

$$\left\|p^M\right\| = \|p\|^M < \frac{1}{2}$$

for some positive integer M. Now p^M and q^N are relatively prime, there exist two integers m and n such that

$$mp^M + nq^N = 1.$$

Then

$$1 = \|1\| = \left\|mp^M + nq^N\right\| \le \left\|mp^M\right\| + \left\|nq^N\right\| < \frac{1}{2} + \frac{1}{2} = 1,$$

a contradiction. Hence $\|q\| = 1$. Now for any positive integer n with the prime factorization

$$n = p_1^{\alpha_1} p_2^{\alpha_2} \cdots p_k^{\alpha_k},$$

one has $\|n\| = \|p\|^{-v_p(n)}$. Thus $\|\cdot\|$ is equivalent to $|\cdot|_p$.

Case II. Suppose that there is a positive integer n such that $\|n\| > 1$. Let n_0 be the least such n. Note that $n_0 \ne 1$, there exists a positive real number α such that $\|n_0\| = n_0^\alpha$ since $\|n_0\| > 1$. With n_0 as a base, we express n as

$$n = a_0 + a_1 n_0 + \cdots + a_k n_0^k, \quad 0 \le a_i < n_0.$$

Then the triangle inequality implies

$$\|n\| \le \|a_0\| + \|a_1 n_0\| + \cdots + \left\|a_k n_0^k\right\|$$
$$\le \|a_0\| + \|a_1\| n_0^\alpha + \cdots + \|a_k\| n_0^{k\alpha}.$$

We have $\|a_i\| \le 1$ since $a_i < n_0$. It follows that

$$\|n\| \le 1 + n_0^\alpha + \cdots + n_0^{k\alpha}$$
$$\le n_0^{k\alpha} \left(1 + n_0^{-\alpha} + \cdots + n_0^{-k\alpha}\right)$$
$$\le n^\alpha \left(1 - n_0^{-\alpha}\right)^{-1}.$$

Thus $\|n\| \leq Cn^\alpha$. Replacing n by n^N, we get $\|n\| \leq C^{1/N}n^\alpha$ and $\|n\| \leq n^\alpha$ as $N \to \infty$.

Next we prove $\|n\| \geq n^\alpha$. With n given as before, then

$$n_0^{k+1} > n \geq n_0^k.$$

We have

$$
\begin{aligned}
\|n\| &\geq \left\|n_0^{k+1}\right\| - \left\|n_0^{k+1} - n\right\| \\
&\geq n_0^{(k+1)\alpha} - \left(n_0^{k+1} - n\right)^\alpha \\
&\geq n_0^{(k+1)\alpha} - \left(n_0^{k+1} - n_0^k\right)^\alpha \qquad \text{(since } n \geq n_0^k) \\
&= n_0^{(k+1)\alpha}\left[1 - \left(1 - n_0^{-1}\right)^\alpha\right] \\
&\geq C'n^\alpha.
\end{aligned}
$$

As before, we replace n by n^N and let $N \to \infty$, it yields $\|n\| \geq n^\alpha$. This proves that $\|n\| = n^\alpha$, and therefore $\|\cdot\|$ is equivalent to $|\cdot|_\infty$. $\qquad \square$

The p-adic valuation defined as above can be extended to elements in $\mathbb{Z}_p$. For any nonzero element x in $\mathbb{Z}_p$, we define

$$v_p(x) = n$$

if $x = p^n u$, $u \in U$.

Proposition A.2.4. *The p-adic ring $\mathbb{Z}_p$ is a complete metric space in which $\mathbb{Z}$ is dense in $\mathbb{Z}_p$.*

Proof. For $x, y \in \mathbb{Z}_p$, we define the distance $d(x, y)$ by

$$d(x, y) = p^{-v_p(x-y)}.$$

Then d is a metric in $\mathbb{Z}_p$. Let $y_j = (y_{1j}, y_{2j}, \ldots, y_{nj}, \ldots)$, $j = 1, 2, \ldots, m, \ldots$, be a Cauchy sequence of $\mathbb{Z}_p$. Then given $M > 0$, there exists a positive integer N such that

$$v_p(y_m - y_n) \geq M$$

for all $m, n \geq N$. It follows that

$$y_{km} \equiv y_{kn} \pmod{p^k}$$

for $1 \leq k \leq [M]$ and $m, n \geq N$. Define $y = (y_n)$ with $y_n = \lim_{j \to \infty} y_{nj}$. Then $y \in \mathbb{Z}_p$ and $\lim_{j \to \infty} y_j = y$. So $\mathbb{Z}_p$ is a complete metric space.

Finally, if $x = (x_1, x_2, \ldots, x_n, \ldots) \in \mathbb{Z}_p$ and $y_n \in \mathbb{Z}$ such that $y_n \equiv x_n$ (mod p^n), then $\lim_{n \to \infty} y_n = x$. This proves that $\mathbb{Z}$ is dense in $\mathbb{Z}_p$. $\square$

Remark A.2.5. The p-adic ring $\mathbb{Z}_p$ is compact since it is contained in the unit ball and is complete.

A.3. The Field of p-Adic Numbers

The quotient field of the p-adic ring is called the field of p-adic numbers and denoted by $\mathbb{Q}_p$. The following facts can be seen immediately.

(a) $\mathbb{Q}_p = \mathbb{Z}_p(p^{-1})$.

(b) Every nonzero element $x \in \mathbb{Q}_p$ can be written as $x = p^n u$ with $u \in U$ and $n \in \mathbb{Z}$.

(c) The p-adic valuation on $\mathbb{Q}$ can be extended to $\mathbb{Q}_p$ as $v_p(x) = n$ if $x = p^n u$, $u \in U$.

(d) The field $\mathbb{Q}_p$ with the metric $d(x, y) = p^{-v_p(x-y)}$ is a locally compact topological space and $\mathbb{Q}$ is dense in $\mathbb{Q}_p$.

For any prime number p and arbitrary integer a_n, the series

$$\sum_{i=0}^{\infty} a_i p^i$$

is convergent in $\mathbb{Z}_p$ since

$$v_p \left(\sum_{i=n}^{\infty} a_i p^i \right) \geq v_p \left(a_n p^n \right) \geq n.$$

As $\mathbb{Z}_p$ is complete, it defines an element of $\mathbb{Z}_p$. Conversely, we shall prove that every $x \in \mathbb{Z}_p$ has a unique representation as a series

$$x = \sum_{i=m}^{\infty} a_i p^i, \quad 0 \leq a_i < p.$$

Proposition A.3.1. *Every $x \in \mathbb{Z}_p$ has a unique representation as a series*

$$x = \sum_{i=m}^{\infty} a_i p^i, \quad 0 \le a_i < p$$

and $m \ge 0$.

Proof. For $x = 0$, we have the representation $0 = \sum_{i=0}^{\infty} 0 \cdot p^i$. Suppose $x = (x_1, x_2, \ldots, x_n, \ldots) \ne 0$ and $v_p(x) = m$. Then $x_k \equiv 0 \pmod{p^m}$ for $k \le m + 1$. Let $y_k = p^{-m} x_k$. Then

$$p^{-m} x = (y_1, y_2, \ldots, y_n, \ldots)$$

and $v_p(y) = 0$. It follows that $y_1 \ne 0$ in $\mathbb{Z}/p\mathbb{Z}$. Now choose a_m such that $0 < a_m < p$ and $a_m \equiv y_1 \pmod{p}$. Then

$$x = a_m p^m + \xi, \quad \xi \in \mathbb{Z}_p, \quad v_p(\xi) \ge m + 1.$$

Continuing this process, we get the representation as asserted.

Suppose that x has two series representations

$$x = \sum_{i=m}^{\infty} a_i p^i = \sum_{i=m'}^{\infty} a_i' p^i, \quad 0 \le a_i, a_i' < p$$

with $a_m \ne 0$ and $a_{m'}' \ne 0$. Then $m = m' = v_p(x)$ and $a_m = a_m'$ by our construction. Hence

$$\sum_{i=m+1}^{\infty} a_i p^i = \sum_{i=m+1}^{\infty} a_i' p^i.$$

Repeating the similar argument, we conclude that $a_i = a_i'$ for all $i \ge m$. $\square$

Corollary A.3.2. *Any element x in $\mathbb{Q}_p$ has the representation*

$$x = \sum_{i=m}^{\infty} a_i p^i, \quad 0 \le a_i < p, \quad m \in \mathbb{Z}.$$

Corollary A.3.3. *The element $\sum_{i=0}^{\infty} a_i p^i$ is a unit of $\mathbb{Z}_p$ if $a_0 \ne 0$.*

A.4. Kummer's Congruences

The well-known Bernoulli numbers B_n, $n = 0, 1, 2, \ldots$, are defined by

$$\frac{t}{e^t - 1} = \sum_{n=0}^{\infty} \frac{B_n t^n}{n!}, \quad |t| < 2\pi.$$

The classical Kummer's congruences asserted congruence relations among Bernoulli numbers modulo prime numbers, i.e., for any prime number p, one has

$$\frac{B_{m+p-1}}{m+p-1} \equiv \frac{B_m}{m} \pmod{p}. \tag{A.4.1}$$

Here m is a positive even integer and p is a prime number such that $p-1$ is not a divisor of m. These congruence relations can be explained in p-adic language and have the general form as follows:

$$m_1 \equiv m_2 \pmod{(p-1)p^n}$$
$$\implies \left(1 - p^{m_1-1}\right) \frac{B_{m_1}}{m_1} \equiv \left(1 - p^{m_2-1}\right) \frac{B_{m_2}}{m_2} \pmod{p^{n+1}}. \tag{A.4.2}$$

Here m_1 is an even integer and p is a prime number such that $p-1$ is not a divisor of m_1.

For any fixed prime number p, if we define the zeta function $\zeta_p^*(s)$ as

$$\zeta_p^*(s) = \sum_{\substack{m=1 \\ (m,p)=1}}^{\infty} m^{-s}, \quad \mathrm{Re}\, s > 1.$$

Then for $\mathrm{Re}\, s > 1$, $\zeta_p^*(s)$ has the Euler product

$$\zeta_p^*(s) = \prod_{q \neq p} \left(1 - q^{-s}\right)^{-1}$$

and it is equal to $(1 - p^{-s})\,\zeta(s)$. Here $\zeta(s)$ is the well-known Riemann zeta function.

Like the Riemann zeta function $\zeta(s)$, $\zeta_p^*(s)$ has a meromorphic analytic continuation in the whole complex plane. Furthermore, for any positive even integer m, one has

$$\zeta_p^*(1 - m) = -\left(1 - p^{m-1}\right) \frac{B_m}{m}.$$

Consequently, the Kummer's congruences in (A.4.2) can be restated as

$$m_1 \equiv m_2 \quad (\mathrm{mod}\ \varphi(p^{n+1})) \implies \zeta_p^*(1 - m_1) \equiv \zeta_p^*(1 - m_2) \quad (\mathrm{mod}\ p^{n+1}).$$

Here the Euler's φ-function is defined by $\varphi(n) = n \prod_{p|n} \left(1 - p^{-1}\right)$.

If we define the zeta function $\zeta_n^*(s)$ as

$$\zeta_n^*(s) = \sum_{\substack{m=1 \\ (m,n)=1}}^{\infty} m^{-s}, \quad \mathrm{Re}\, s > 1,$$

and denote its analytic continuation in the whole complex plane by the same notation. Then we have the following generalization of the Kummer's congruences which we shall prove by considering integrations over the inverse limit of $\mathbb{Z}/np^N\mathbb{Z}$ (see Chapter II of [47] for the definition) with respect to a certain p-adic measure.

Theorem A.4.1. *If n is a positive integer and m_1, m_2 are positive even integers such that the following conditions:*

(a) $m_1 \equiv m_2 \ (\mathrm{mod}\ \varphi(n))$;

(b) *$p - 1$ is not a divisor of m_1 for any prime divisor p of n.*

Then we have

$$\zeta_n^*(1 - m_1) \equiv \zeta_n^*(1 - m_2) \quad (\mathrm{mod}\ n)$$

or equivalently that

$$\frac{B_{m_1}}{m_1} \prod_{p|n} \left(1 - p^{m_1-1}\right) \equiv \frac{B_{m_2}}{m_2} \prod_{p|n} \left(1 - p^{m_2-1}\right) \quad (\mathrm{mod}\ n).$$

We shall express $\zeta_n^*(1 - m)$ as an integral of x^{m-1} over a certain space with respect to a p-adic measure. With the help of the Fermat-Euler's congruence relation

$$(a, n) = 1 \implies a^{\varphi(n)} \equiv 1 \quad (\mathrm{mod}\ n),$$

we conclude our assertion in the theorem.

Let p be a prime number. The sets $\mathbb{Z}_p$ and $\mathbb{Q}_p$ are the ring of p-adic integers and the field of p-adic numbers, respectively. Let Ω_p be the algebraic completion of $\mathbb{Q}_p$. For a fixed positive integer n, we let X_n be the inverse limit of $\mathbb{Z}/np^N\mathbb{Z}$, i.e.,

$$X_n = \varprojlim \mathbb{Z}/np^N\mathbb{Z},$$

where the map from $\mathbb{Z}/np^M\mathbb{Z}$ to $\mathbb{Z}/np^N\mathbb{Z}$ for $M \geq N$ is the reduction modulo np^N. Denote by $a + np^N\mathbb{Z}_p$ the set of x in X_n which map to a in $\mathbb{Z}/np^N\mathbb{Z}$ under the natural projection map from X_n to $\mathbb{Z}/np^N\mathbb{Z}$.

Fix an rth root of unity ϵ with r relatively prime to n. Also suppose that ϵ is not a p^nth root of unity for any N. Define

$$\mu_\epsilon\left(a + np^N\mathbb{Z}_p\right) = \frac{\epsilon^a}{1 - \epsilon^{np^N}}$$

and

$$\mu\left(a + np^N\mathbb{Z}_p\right) = \sum_{\substack{\epsilon^r = 1 \\ \epsilon \neq 1}} \mu_\epsilon\left(a + np^N\mathbb{Z}_p\right).$$

The above p-adic measure was given in [38] and it is also known as Mazure measure.

Note that

$$X_n = \bigcup_{0 \leq a < n} a + n\mathbb{Z}_p$$

is a disjoint union of n topological spaces isomorphic to $\mathbb{Z}_p$. We also have

$$a + np^N\mathbb{Z}_p = \bigcup_{0 \leq b < p} (a + bnp^N) + np^{N+1}\mathbb{Z}_p.$$

The above is a disjoint union of p compact open sets. It is easy to verify that

$$\mu\left(a + np^N\mathbb{Z}_p\right) = \sum_{b=0}^{p-1} \mu\left((a + bnp^N) + np^{N+1}\mathbb{Z}_p\right).$$

For any continuous function $f\colon X_n \to \Omega_p$, we define

$$\int_{X_n} f(x)\,d\mu(x) = \lim_{N \to \infty} \sum_{0 \leq a < np^N} f(a)\mu\left(a + np^N\mathbb{Z}_p\right).$$

Proposition A.4.2. *For any positive integers m and n, we have*

$$\int_{X_n} x^{m-1}\, d\mu(x) = (1 - r^m)\frac{B_m}{m}.$$

Proof. For each t in Ω_p with $v_p(t) > 1/(p-1)$, the exponential function e^{tx} defined by the power series

$$e^{tx} = \sum_{i=0}^{\infty} \frac{t^i x^i}{i!}$$

is a continuous function on X_n. Hence we have

$$\int_{X_n} e^{tx}\, d\mu_\epsilon(x) = \lim_{N \to \infty} \frac{1}{1 - \epsilon^{np^N}} \sum_{0 \le a < np^N} \epsilon^a e^{ta}$$

$$= \lim_{N \to \infty} \frac{1 - \epsilon^{np^N} e^{np^N t}}{\left(1 - \epsilon^{np^N}\right)\left(1 - \epsilon e^t\right)}$$

$$= \frac{1}{1 - \epsilon e^t}$$

since $e^{np^N t} \to 1$ as $N \to \infty$. Consequently, we get

$$\int_{X_n} e^{tx}\, d\mu_\epsilon(x) = \frac{1}{1 - \epsilon e^t}$$

$$= \frac{1 + \epsilon e^t + \cdots + \epsilon^{r-1} e^{(r-1)t}}{1 - e^{rt}}.$$

Letting ϵ range over all rth roots of unity except 1 and summing together, we get

$$\int_{X_n} e^{tx}\, d\mu(x) = \frac{r - \left(1 + e^t + \cdots + e^{(r-1)t}\right)}{1 - e^{rt}}$$

$$= \frac{r}{1 - e^{rt}} - \frac{1}{1 - e^t}$$

$$= \sum_{m=1}^{\infty} \frac{(1 - r^m) B_m t^{m-1}}{m!}.$$

By comparing the coefficients of t, we get our formula as asserted. $\qquad\square$

Remark A.4.3. In order to determine the radius of convergence of the power series of exponential function defined by

$$e^x = \sum_{n=0}^{\infty} \frac{x^n}{n!},$$

we have to find $v_p(n!)$. By Exercise A.2.2, we have

$$v_p(n!) = \frac{n - S_n}{p - 1}.$$

Note that $S_n \leq (p - 1)(\log_p n + 1)$, it follows that $v_p(n!)$ is approximately equal to $n/(p - 1)$ and so

$$v_p\left(\frac{x^n}{n!}\right) \to \infty \quad \Longleftrightarrow \quad v_p(x) > \frac{1}{p - 1}.$$

Proposition A.4.4. *Let X_n^* be elements of X_n which map onto $(\mathbb{Z}/n\mathbb{Z})^*$, the invertible elements of $\mathbb{Z}/n\mathbb{Z}$. Then*

$$\int_{X_n^*} x^{m-1}\, d\mu(x) = (1 - r^m) \prod_{p|n} \left(1 - p^{m-1}\right) \frac{B_m}{m}.$$

Proof. By the inclusion-exclusion principle, we decompose the integration into the following:

$$\int_{X_n^*} = \int_{X_n} - \sum_{p_j|n} \int_{p_j X_n} + \sum_{p_i,p_j|n} \int_{p_i p_j X_n} + \cdots + (-1)^k \int_{p_1 p_2 \cdots p_k X_n}.$$

Here $p_1, p_2, \ldots, p_k$ are distinct prime divisors of n. To prove the proposition, it suffices to prove that

$$\int_{\alpha X_n} x^{m-1}\, d\mu(x) = (1 - r^m)\, \alpha^{m-1} \frac{B_m}{m}$$

for any integer α which is a prime divisor or a product of distinct prime divisors of n.

Again, we consider the integration of e^{tx}:

$$\int_{\alpha X_n} e^{tx}\, d\mu_\epsilon(x) = \lim_{N \to \infty} \frac{1}{1 - \epsilon^{np^N}} \sum_{0 \leq b < np^N/\alpha} \left(\epsilon e^t\right)^{\alpha b}$$

$$= \lim_{N \to \infty} \frac{1 - \left(\epsilon e^t\right)^{np^N}}{\left(1 - \epsilon^{np^N}\right)\left(1 - \epsilon^\alpha e^{\alpha t}\right)}$$

$$= \frac{1}{1 - \epsilon^\alpha e^{\alpha t}}.$$

Since r is relatively prime to α, the mapping ϵ to ϵ^α causes a permutation among rth roots of unity. Hence

$$\int_{\alpha X_n} e^{tx}\, d\mu(x) = \frac{r - \left(1 + e^{\alpha t} + \cdots + e^{(r-1)\alpha t}\right)}{1 - e^{r\alpha t}}$$

$$= \frac{r}{1 - e^{r\alpha t}} - \frac{1}{1 - e^{\alpha t}}$$

$$= \sum_{m=1}^{\infty} \frac{(1 - r^m)\, B_m(\alpha t)^{m-1}}{m!}.$$

By equating the coefficients of t, we get our assertion. $\qquad\square$

Now we are ready to prove our main theorem.

Proof of Theorem A.4.1. For any element x in $(\mathbb{Z}/n\mathbb{Z})^*$, we have the congruence relation

$$x^{m_1-1} \equiv x^{m_2-1} \pmod{n}$$

since $m_1 - m_2$ is a multiple of $\varphi(n)$. Hence for any prime divisor p of n, with the p-adic measure $\mu(x)$ defined on X_n, we have

$$\int_{X_n^*} x^{m_1-1}\, d\mu(x) \equiv \int_{X_n^*} x^{m_2-1}\, d\mu(x) \pmod{p^\alpha},$$

where $\alpha = v_p(n)$ is the highest power of p dividing n. On the other hand, we have $r^{m_1} - 1 \in (\mathbb{Z}/n\mathbb{Z})^*$ since $r^{m_1} - 1 \in (\mathbb{Z}/p^\alpha\mathbb{Z})^*$ for any prime divisor p of n and α is the highest power of p dividing n. Also we have

$$r^{m_1} - 1 \equiv r^{m_2} - 1 \pmod{n}$$

since $(r, n) = 1$ and $m_1 \equiv m_2 \pmod{\varphi(n)}$. Hence

$$\frac{1}{1 - r^{m_1}} \int_{X_n^*} x^{m_1-1}\, d\mu(x) \equiv \frac{1}{1 - r^{m_2}} \int_{X_n^*} x^{m_2-1}\, d\mu(x) \pmod{p^\alpha}.$$

This is equivalent to

$$\zeta_n^*(1 - m_1) \equiv \zeta_n^*(1 - m_2) \pmod{p^\alpha},$$

and thus

$$\zeta_n^*(1 - m_1) \equiv \zeta_n^*(1 - m_2) \pmod{n}.$$

$\qquad\square$

Remark A.4.5. When $n = p^\alpha$ is a prime power, we have

$$\int_{X_n^*} x^{m-1}\, d\mu(x) = \int_{\mathbb{Z}_p} x^{m-1}\, d\mu(x) - \int_{p\mathbb{Z}_p} x^{m-1}\, d\mu(x)$$

$$= \left(1 - p^{m-1}\right)\left(1 - r^m\right)\frac{B_m}{m}.$$

This is the same as the formula

$$\int_{\mathbb{Z}_p^*} x^{m-1}\, d\mu(x) = \left(1 - p^{m-1}\right)\left(1 - r^m\right)\frac{B_m}{m}$$

which was obtained in [38].

Appendix B

Weighted Sum Formulae of Multiple Zeta Values

B.1. Introduction and Notations

For a pair of positive integers p and q with $q \geq 2$, the classical Euler sum is defined as [7, 9, 11, 29, 30, 34, 35, 41]

$$S_{p,q} = \sum_{k=1}^{\infty} \frac{1}{k^q} \sum_{j=1}^{k} \frac{1}{j^p}.$$

Multiple zeta values [23, 25, 44, 52] are natural generalizations of Euler sums. For positive integers $\alpha_1, \alpha_2, \ldots, \alpha_r$ with $\alpha_r \geq 2$, the multiple zeta values or r-fold Euler sum is defined as

$$\zeta(\alpha_1, \alpha_2, \ldots, \alpha_r) = \sum_{k_r=1}^{\infty} \frac{1}{k_r^{\alpha_r}} \sum_{k_{r-1}=1}^{k_r-1} \frac{1}{k_{r-1}^{\alpha_{r-1}}} \cdots \sum_{k_1=1}^{k_2-1} \frac{1}{k_1^{\alpha_1}}$$

or equivalently as

$$\sum_{1 \leq n_1 < n_2 < \cdots < n_r} n_1^{-\alpha_1} n_2^{-\alpha_2} \cdots n_r^{-\alpha_r}.$$

Here the numbers r and $|\alpha| = \alpha_1 + \alpha_2 + \cdots + \alpha_r$ are the depth and the weight of $\zeta(\alpha_1, \alpha_2, \ldots, \alpha_r)$, respectively. For our convenience, we let $\{1\}^k$

be k repetitions of 1. For example,

$$\zeta(\{1\}^3, 3) = \zeta(1,1,1,3) \quad \text{and} \quad \zeta(\{1\}^4, 3) = \zeta(1,1,1,1,3).$$

For positive integers m and n, multiple zeta values of the form

$$\zeta(\{1\}^{m-1}, n+1)$$

were fully investigated and substantial results were obtained [11, 35, 41, 44]. On the one hand, they can be expressed in terms of special values at positive integers of Riemann zeta function. On the other hand, they can be expressed as Drinfeld integrals [29]

$$\zeta(\{1\}^{m-1}, n+1) = \int_{0<t_1<t_2<\cdots<t_{m+n}<1} \prod_{j=1}^{m} \frac{dt_j}{1-t_j} \prod_{k=m+1}^{m+n} \frac{dt_k}{t_k}.$$

An elementary consideration then leads to integral representation

$$\zeta(\{1\}^{m-1}, n+1) = \frac{1}{(m-1)!n!} \int_0^1 \left(\log \frac{1}{1-t}\right)^{m-1} \left(\log \frac{1}{t}\right)^n \frac{dt}{1-t}.$$

Therefore, it is easy to express the product of two multiple zeta values of the form $\zeta(\{1\}^{m-1}, n+1)$ as a double integral over the square $[0,1] \times [0,1]$. Bisecting the domain into two triangular regions and expressing the integrals as sums of multiple zeta values, we get the following

Theorem B.1.1. *For positive integers m, n, p and q, we have*

$$\sum_{n_1+n_2=n} \sum_{p_1+p_2=p-1} \binom{m-1+p_1}{m-1}\binom{q+n_2}{q}$$

$$\times \sum_{|\alpha|=n_1+p_2+1} \zeta(\{1\}^{m-1+p_1}, \alpha_1, \ldots, \alpha_{p_2+1}, q+n_2+1)$$

$$+ \sum_{q_1+q_2=q} \sum_{m_1+m_2=m-1} \binom{p-1+m_1}{p-1}\binom{n+q_2}{n}$$

$$\times \sum_{|\beta|=q_1+m_2+1} \zeta(\{1\}^{p-1+m_1}, \beta_1, \ldots, \beta_{m_2+1}, n+q_2+1)$$

$$= \zeta(\{1\}^{m-1}, n+1)\zeta(\{1\}^{p-1}, q+1).$$

The following proposition first given in [29] has its original from Ohno's paper [44]. It is a basic tool to express a single or a sum of multiple zeta values as a double integral and vice versa.

Proposition B.1.2. *For a nonnegative integer p and positive integers q, m and n with $m \geq q$, we have*

$$\sum_{|\alpha|=m} \zeta(\{1\}^p, \alpha_1, \alpha_2, \ldots, \alpha_q + n)$$

$$= C(p,q,m,n) \int_{0<t_1<t_2<1} \left(\log \frac{1}{1-t_1}\right)^p \left(\log \frac{t_2}{t_1}\right)^{m-q} \left(\log \frac{1-t_1}{1-t_2}\right)^{q-1}$$

$$\times \left(\log \frac{1}{t_2}\right)^{n-1} \frac{dt_1 dt_2}{(1-t_1)t_2},$$

where $C(p,q,m,n) = \{p!(q-1)!(m-q)!(n-1)!\}^{-1}$.

The *sum formula* [11, 29, 32]

$$\sum_{|\alpha|=m} \zeta(\alpha_1, \alpha_2, \ldots, \alpha_r + 1) = \zeta(m+1)$$

along with Proposition B.1.2 enable us to express $\zeta(m+1)$ as a double integral over a simplex in $\mathbb{R}^2$. So if we express $\zeta(m+1)\zeta(p+1)$ as a triple integral, then the decomposition of the domain into standard simplices of the form

$$(t_1, t_2, t_3) \in [0,1]^3, \quad t_1 < t_2 < t_3$$

leads to the following

Theorem B.1.3. *For positive integers m, p and r with $m \geq r$, we have*

$$\sum_{|\alpha|=m} \zeta(\alpha_1, \alpha_2, \ldots, \alpha_r + 1, p + 1)$$

$$+ \sum_{p_1+p_2+p_3=p} \binom{m-r+p_2}{m-r} \sum_{|\alpha|=m+p_2} \zeta(p_1+1, \alpha_1, \alpha_2, \ldots, \alpha_r + p_3 + 1)$$

$$+ \sum_{r_1+r_2=r-1} \sum_{p_1+p_2=p} \sum_{\ell_1+\ell_2=m-r} \binom{p_1+\ell_2}{p_1} \sum_{|\alpha|=r_1+\ell_1+1}$$

$$\times \sum_{|\beta|=r_2+p_1+\ell_2+1} \zeta(\alpha_1, \alpha_2, \ldots, \alpha_{r_1}+1, \beta_1, \beta_2, \ldots, \beta_{r_2}+1 + p_2 + 1)$$

$$= \zeta(m+1)\zeta(p+1).$$

Remark B.1.4. Setting $m = p = 1$ in Theorem B.1.1, we get

$$\zeta(n+1)\zeta(q+1) = \sum_{n_1+n_2=n} \binom{q+n_2}{q}\zeta(n_1+1, q+n_2+1)$$

$$+ \sum_{q_1+q_2=q} \binom{n+q_2}{n}\zeta(q_1+1, n+q_2+1).$$

Such a formula can be generalized to the product of a finite number of Riemann zeta values. We shall produce the generalization in our final section.

B.2. The Proof of Theorem B.1.1 and Its Consequences

The evaluations of multiple zeta values of the form $\zeta(\{1\}^{m-1}, n+1)$ are given in [30] as follows:

$$\zeta(\{1\}^{2k}, 2n) = \zeta(2n+2k)$$

$$+ \sum_{p=0}^{k-1}\sum_{\alpha=2}^{2n-1}(-1)^{\alpha+p+1}\zeta(\{1\}^p, 2n+1-\alpha)\zeta(\{1\}^{2k-p-1}, \alpha),$$

$$\zeta(\{1\}^{2k+1}, 2n+1) = -\zeta(2n+2k+2)$$

$$+ \frac{1}{2}\sum_{p=0}^{2k}\sum_{\alpha=2}^{2n}(-1)^{\alpha+p}\zeta(\{1\}^p, 2n+2-\alpha)\zeta(\{1\}^{2k-p}, \alpha),$$

$$(2k+1)\zeta(\{1\}^{2k}, 2n+1)$$

$$= -(2k+1)\zeta(2n+2k+1) - 2\sum_{\alpha=1}^{2k}\zeta(\alpha, 2n+2k+1-\alpha)$$

$$+ \sum_{p=0}^{k-1}(-1)^p(2k-2p-1)\sum_{\alpha=2}^{2n}(-1)^{\alpha}\zeta(\{1\}^p, 2n+2-\alpha)\zeta(\{1\}^{2k-p-1}, \alpha)$$

and

$$(k+1)\zeta(\{1\}^{2k+1}, 2n)$$

$$= k\zeta(2n+2k+1) + \sum_{\alpha=1}^{2k+1} \zeta(\alpha, 2n+2k+1-\alpha)$$

$$+ \sum_{p=0}^{k-1}(-1)^p(k-p)\sum_{\alpha=2}^{2n-1}(-1)^{\alpha+1}\zeta(\{1\}^p, 2n+1-\alpha)\zeta(\{1\}^{2k-p}, \alpha).$$

In particular, all of them can be expressed in terms of single zeta values recursively.

The Drinfeld integral for $\zeta(\{1\}^{m-1}, n+1)$ is given by [11, 29, 44]

$$\int_{0<t_1<t_2<\cdots<t_{m+n}<1} \prod_{j=1}^{m} \frac{dt_j}{1-t_j} \prod_{k=m+1}^{m+n} \frac{dt_k}{t_k}.$$

For any permutation σ of the set $S = \{1, 2, \ldots, m-1\}$ and permutation τ of the set $T = \{m+1, m+2, \ldots, m+n\}$, let $D_{\sigma,\tau}$ be the simplex defined by

$$0 < t_{\sigma(1)} < t_{\sigma(2)} < \cdots < t_{\sigma(m-1)} < t_m < t_{\tau(m+1)} < \cdots < t_{\tau(m+n)} < 1.$$

Then the integral

$$\int_{D_{\sigma,\tau}} \prod_{j=1}^{m} \frac{dt_j}{1-t_j} \prod_{k=m+1}^{m+n} \frac{dt_k}{t_k}$$

has the same value as $\zeta(\{1\}^{m-1}, n+1)$, so that

$$\zeta(\{1\}^{m-1}, n+1) = \frac{1}{(m-1)!n!}\int_0^1 \frac{dt}{1-t} \prod_{j=1}^{m-1}\int_0^t \frac{dt_j}{1-t_j} \prod_{k=m+1}^{m+n}\int_t^1 \frac{dt_k}{t_k}$$

$$= \frac{1}{(m-1)!n!}\int_0^1 \left(\log\frac{1}{1-t}\right)^{m-1}\left(\log\frac{1}{t}\right)^n \frac{dt}{1-t}.$$
$$\tag{B.2.1}$$

Now we are ready to prove Theorem B.1.1.

Proof of Theorem B.1.1. First we rewrite the product

$$\zeta(\{1\}^{m-1}, n+1)\zeta(\{1\}^{p-1}, q+1)$$

as a double integral over $[0,1] \times [0,1]$ by (B.2.1), namely,

$$\zeta(\{1\}^{m-1}, n+1)\zeta(\{1\}^{p-1}, q+1) = \int_0^1 \int_0^1 F(t,u) \frac{dtdu}{(1-t)(1-u)},$$

where

$$F(t,u) = \frac{1}{(m-1)!n!(p-1)!q!} \left(\log \frac{1}{1-t}\right)^{m-1} \left(\log \frac{1}{t}\right)^n$$

$$\times \left(\log \frac{1}{1-u}\right)^{p-1} \left(\log \frac{1}{u}\right)^q.$$

Decompose the square $[0,1] \times [0,1]$ into two halves defined by

$$D_1 : 0 < t < u < 1 \quad \text{and} \quad D_2 : 0 < u < t < 1,$$

respectively. When in D_1, we replace the factor $\frac{1}{n!}\left(\log \frac{1}{t}\right)^n$ in $F(t,u)$ by its binomial expansion

$$\sum_{n_1+n_2=n} \frac{1}{n_1!n_2!} \left(\log \frac{u}{t}\right)^{n_1} \left(\log \frac{1}{u}\right)^{n_2}$$

and the other factor $\frac{1}{(p-1)!}\left(\log \frac{1}{1-u}\right)^{p-1}$ in $F(t,u)$ by

$$\sum_{p_1+p_2=p-1} \frac{1}{p_1!p_2!} \left(\log \frac{1}{1-t}\right)^{p_1} \left(\log \frac{1-t}{1-u}\right)^{p_2}.$$

Consequently, we have

$$\iint_{D_1} F(t,u) \frac{dtdu}{(1-t)(1-u)}$$

$$= \sum_{n_1+n_2=n} \sum_{p_1+p_2=p-1} \frac{1}{(m-1)!n_1!n_2!p_1!p_2!q!} \int_{0<t<u<1} \left(\log \frac{1}{1-t}\right)^{m-1+p_1}$$

$$\times \left(\log \frac{1-t}{1-u}\right)^{p_2} \left(\log \frac{u}{t}\right)^{n_1} \left(\log \frac{1}{u}\right)^{q+n_2} \frac{dtdu}{(1-t)(1-u)}$$

which is equal to

$$\sum_{n_1+n_2=n} \sum_{p_1+p_2=p-1} \binom{m-1+p_1}{m-1} \binom{q+n_2}{q}$$

$$\times \sum_{|\alpha|=p_2+n_1+1} \zeta(\{1\}^{m-1+p_1}, \alpha_1, \ldots, \alpha_{p_2+1}, q+n_2+1).$$

By symmetry, we express the integration over the domain D_2 as another sum of multiple zeta values after a similar replacement of $F(t,u)$. $\qquad \square$

Here we list a few immediate consequences of Theorem B.1.1.

1. When $m = p = 1$, we have for positive integers n and q,

$$\zeta(n+1)\zeta(q+1) = \sum_{n_1+n_2=n} \binom{q+n_2}{q} \zeta(n_1+1, q+n_2+1)$$

$$+ \sum_{q_1+q_2=q} \binom{n+q_2}{n} \zeta(q_1+1, n+q_2+1).$$

After a rearrangement with the reflection formula

$$\zeta(p,q) + \zeta(q,p) = \zeta(p)\zeta(q) - \zeta(p+q),$$

we get for $w = n+q+2$ that

$$\sum_{\alpha=1}^{n} \binom{w-\alpha-1}{q} \zeta(\alpha, w-\alpha) + \sum_{\alpha=1}^{q} \binom{w-\alpha-1}{n} \zeta(\alpha, w-\alpha) = \zeta(w)$$

which is different from the sum formula due to Euler or the relation obtained from the partial fractions [41].

2. When $n = p = 1$, we have for positive integers m and q that

$$\zeta(m+1)\zeta(q+1) = (q+1)\zeta(\{1\}^m, q+2) + \zeta(\{1\}^{m-1}, 2, q+1)$$

$$+ \sum_{q_1+q_2=q} \sum_{m_1+m_2=m-1} (q_2+1)$$

$$\times \sum_{|\mathbf{c}|=m_2+q_1+1} \zeta(\{1\}^{m_1}, c_1, \ldots, c_{m_2+1}, q_2+2).$$

3. When $p = q = 1$, we get for positive integers m and n that

$$\zeta(2)\zeta(\{1\}^{m-1}, n+1)$$

$$= \sum_{n_1+n_2=n} (n_2+1)\zeta(\{1\}^{m-1}, n_1+1, n_2+2)$$

$$+ m(n+1)\zeta(\{1\}^m, n+2)$$

$$+ \sum_{m_1+m_2=m-1} \sum_{|\alpha|=m_2+2} \zeta(\{1\}^{m_1}, \alpha_1, \ldots, \alpha_{m_2+1}, n+1).$$

B.3. The Proof of Theorem B.1.3 and Generalizations

For positive integers m and r with $m \geq r$, the sum formula [29, 32, 44] asserted that

$$\sum_{|\alpha|=m} \zeta(\alpha_1, \alpha_2, \ldots, \alpha_r + 1) = \zeta(m + 1).$$

However, by Proposition B.1.2, the left-hand side on the above is equal to the double integral

$$\frac{1}{(r-1)!(m-r)!} \int_{0<t_1<t_2<1} \left(\log \frac{1-t_1}{1-t_2} \right)^{r-1} \left(\log \frac{t_2}{t_1} \right)^{m-r} \frac{dt_1 dt_2}{(1-t_1)t_2}.$$

Therefore, we conclude immediately that

$$\int_{0<t_1<t_2<1} \int_0^1 G(t_1, t_2, u) \frac{dt_1 dt_2 du}{(1-t_1)t_2(1-u)} = \zeta(m+1)\zeta(p+1),$$

where

$$G(t_1, t_2, u) = \frac{1}{(r-1)!(m-r)!p!} \left(\log \frac{1-t_1}{1-t_2} \right)^{r-1} \left(\log \frac{t_2}{t_1} \right)^{m-r} \left(\log \frac{1}{u} \right)^p.$$

Proof of Theorem B.1.3. Notations as above, we decompose the domain of integration into three simplices as follows:

$$D_1 : 0 < t_1 < t_2 < u < 1, \quad D_2 : 0 < u < t_1 < t_2 < 1$$

and

$$D_3 : 0 < t_1 < u < t_2 < 1.$$

It is immediately, by Proposition B.1.2, that

$$\int_{D_1} G(t_1, t_2, u) \frac{dt_1 dt_2 du}{(1-t_1)t_2(1-u)} = \sum_{|\alpha|=m} \zeta(\alpha_1, \alpha_2, \ldots, \alpha_r + 1, p + 1).$$

In the second domain $D_2 : 0 < u < t_1 < t_2 < 1$, we have to replace the factor $\frac{1}{p!} \left(\log \frac{1}{u} \right)^p$ in $G(t_1, t_2, u)$ by its multinomial expansion

$$\sum_{p_1+p_2+p_3=p} \frac{1}{p_1! p_2! p_3!} \left(\log \frac{t_1}{u} \right)^{p_1} \left(\log \frac{t_2}{t_1} \right)^{p_2} \left(\log \frac{1}{t_2} \right)^{p_3}$$

so that the integration of $G(t_1, t_2, u)$ over D_2 is equal to

$$\sum_{p_1+p_2+p_3=p} \binom{m-r+p_2}{m-r} \sum_{|\alpha|=m+p_2} \zeta(p_1+1, \alpha_1, \alpha_2, \ldots, \alpha_r + p_3 + 1).$$

Finally, in D_3, we rewrite $G(t_1, t_2, u)$ as

$$\sum_{r_1+r_2=r-1} \sum_{\ell_1+\ell_2=m-r} \sum_{p_1+p_2=p} \frac{1}{r_1! r_2! \ell_1! \ell_2! p_1! p_2!}$$

$$\times \left(\log \frac{1-t_1}{1-u}\right)^{r_1} \left(\log \frac{u}{t_1}\right)^{\ell_1} \left(\log \frac{1-u}{1-t_2}\right)^{r_2} \left(\log \frac{t_2}{u}\right)^{\ell_2+p_1} \left(\log \frac{1}{t_2}\right)^{p_2}$$

so that the integration of $G(t_1, t_2, u)$ over D_3 is equal to

$$\sum_{r_1+r_2=r-1} \sum_{\ell_1+\ell_2=m-r} \sum_{p_1+p_2=p} \binom{p_1+\ell_2}{p_1}$$

$$\times \sum_{|\alpha|=r_1+\ell_1+1} \sum_{|\beta|=r_2+p_1+\ell_2+1} \zeta(\alpha_1, \ldots, \alpha_{r_1+1}, \beta_1, \ldots, \beta_{r_2+1} + p_2 + 1).$$

$\square$

Remark B.3.1. In Theorem B.1.3, if $r = m$, we get

$$\zeta(\{1\}^{m-1}, 2, p+1) + \sum_{p_1+p_2+p_3=p} \sum_{|\alpha|=m+p_2} \zeta(p_1+1, \alpha_1, \ldots, \alpha_m + p_3 + 1)$$

$$+ \sum_{m_1+m_2=m-1} \sum_{p_1+p_2=p} \sum_{|\beta|=m_2+p_1+1} \zeta(\{1\}^{m_1+1}, \beta_1, \ldots, \beta_{m_2+1} + p_2 + 1)$$

$$= \zeta(m+1)\zeta(p+1).$$

On the other hand, if $r = 1$, we get another sum formula

$$\zeta(m+p+2) = \sum_{\substack{p_1+p_2+p_3=p \\ p_1<p}} \binom{m-1+p_2}{m-1} \zeta(p_1+1, m+p_2+p_3+1)$$

$$+ \sum_{p_1+p_2=p} \sum_{m_1+m_2=m-1} \binom{p_1+m_2}{p_1} \zeta(m_1+1, m_2+p+2)$$

after an elimination.

B.4. Generalizations

It is hard to write down the multiple version of Theorem B.1.1. However, it is much easier to express the product of k Riemann zeta values as

$$\prod_{j=1}^{k} \zeta(\alpha_j + 1) = \prod_{j=1}^{k} \frac{1}{\alpha_j!} \int_0^1 \left(\log \frac{1}{t_j} \right)^{\alpha_j} \frac{dt_j}{1 - t_j}.$$

Decompose the cube $[0,1]^k$ into $k!$ simplices of the form

$$D_\sigma : 0 < t_{\sigma(1)} < t_{\sigma(2)} < \cdots < t_{\sigma(k)} < 1,$$

where σ is a permutation of the set $S = \{1, 2, \ldots, k\}$ so that

$$\prod_{j=1}^{k} \zeta(\alpha_j + 1) = \sum_{\sigma \in S_k} \int_{D_\sigma} \prod_{j=1}^{k} \frac{1}{\alpha_j!} \left(\log \frac{1}{t_j} \right)^{\alpha_j} \frac{dt_j}{1 - t_j}.$$

Here S_k is the set of all permutations of k objects. The remaining now is to change the integrand according to D_σ as we have done before. If $\sigma = e$ is an identity, D_e is defined by

$$0 < t_1 < t_2 < \cdots < t_k < 1.$$

By convention, we let $t_{k+1} = 1$. In the domain D_e, we have to replace the factor $\frac{1}{\alpha_j!} \left(\log \frac{1}{t_j} \right)^{\alpha_j}$ by its multinomial expansion

$$\sum_{\alpha_j = \alpha_{jj} + \cdots + \alpha_{jk}} \prod_{\ell=j}^{k} \frac{1}{\alpha_{j\ell}!} \left(\log \frac{t_{\ell+1}}{t_\ell} \right)^{\alpha_{j\ell}}, \quad 1 \le j < k$$

so that the integration gives a sum of multiple zeta values.

If we employ the notation of multinomial coefficient

$$\binom{r_1 + r_2 + \cdots + r_k}{r_1, r_2, \ldots, r_k} = \frac{(r_1 + r_2 + \cdots + r_k)!}{r_1! r_2! \cdots r_k!},$$

then

$$\prod_{j=1}^{k} \zeta(\alpha_j + 1) = \sum_{\sigma \in S_k} G(\sigma).$$

with

$$G(e) = \sum_{\alpha_1 = \alpha_{11} + \alpha_{12} + \cdots + \alpha_{1k}} \sum_{\alpha_2 = \alpha_{22} + \alpha_{23} + \cdots + \alpha_{2k}} \cdots \sum_{\alpha_k = \alpha_{kk}} \binom{\alpha_{12} + \alpha_{22}}{\alpha_{12}, \alpha_{22}}$$

$$\times \binom{\alpha_{13} + \alpha_{23} + \alpha_{33}}{\alpha_{13}, \alpha_{23}, \alpha_{33}} \cdots \binom{\alpha_{1k} + \alpha_{2k} + \cdots + \alpha_{kk}}{\alpha_{1k}, \alpha_{2k}, \ldots, \alpha_{kk}}$$

$$\times \zeta(\alpha_{11} + 1, \alpha_{12} + \alpha_{22} + 1, \ldots, \alpha_{1k} + \alpha_{2k} + \cdots + \alpha_{kk} + 1)$$

which is the value of the integral over the domain $0 < t_1 < t_2 < \cdots < t_k < 1$. Also $G(\sigma)$ is obtained from $G(e)$ by substituting $\alpha_{\sigma(1)}, \alpha_{\sigma(2)}, \ldots, \alpha_{\sigma(k)}$ for $\alpha_1, \alpha_2, \ldots, \alpha_k$, respectively.

There is a *weighted sum formula* in [4] given by

$$(-1)^k \sum_{|\alpha| = p+k} \binom{q + \alpha_k - 1}{q} \zeta(\alpha_1, \alpha_2, \ldots, \alpha_{k-1}, \alpha_k + q)$$

$$+ \sum_{|\alpha| = q+k} \binom{p + \alpha_k - 1}{p} \zeta(\alpha_1, \alpha_2, \ldots, \alpha_{k-1}, \alpha_k + p)$$

$$= \sum_{j=0}^{k-2} (-1)^j \zeta(\{1\}^{p-1}, k - j) \zeta(\{1\}^{q-1}, j + 2).$$

Such an identity follows from the consideration of the double integral

$$\frac{1}{p! q! (k-2)!} \int_0^1 \int_0^1 \left(\log \frac{1}{t} \right)^p \left(\log \frac{1-u}{1-t} \right)^{k-2} \left(\log \frac{1}{u} \right)^q \frac{dt du}{(1-t)(1-u)}. \tag{B.4.1}$$

Replacing the factor $\left(\log \frac{1-u}{1-t} \right)^{k-2}$ in integrand by its binomial expansion

$$\sum_{j=0}^{k-2} (-1)^j \frac{(k-2)!}{j! (k-2-j)!} \left(\log \frac{1}{1-t} \right)^{k-2-j} \left(\log \frac{1}{1-u} \right)^j,$$

we get the value of the integral as

$$\sum_{j=0}^{k-2} (-1)^j \zeta(\{1\}^{k-j-2}, p + 1) \zeta(\{1\}^j, q + 1)$$

which is equal to

$$\sum_{j=0}^{k-2} (-1)^j \zeta(\{1\}^{p-1}, k - j) \zeta(\{1\}^{q-1}, j + 2)$$

by Drinfeld duality theorem

$$\zeta(\{1\}^{m-1}, n+1) = \zeta(\{1\}^{n-1}, m+1).$$

On the other hand, the decomposition of the square $[0,1] \times [0,1]$ into two simplices $D_1 : 0 < t < u < 1$ and $D_2 : 0 < u < t < 1$, then transform the double integral into two sums of multiple zeta values.

Next we employ the same alternating sum to represent another pair of multiple zeta values. For positive integers $n \geq 2$ and nonnegative integers p and q, we define

$$G_n(p, q) = \sum_{k_0=1}^{\infty} \frac{1}{k_0^n} \sum_{k_1=1}^{k_0} \frac{1}{k_1} \cdots \sum_{k_q=1}^{k_{q-1}} \frac{1}{k_q} \left(\sum_{\ell_1=1}^{k_0-1} \frac{1}{\ell_1} \cdots \sum_{\ell_p=1}^{\ell_{p-1}-1} \frac{1}{\ell_p} \right).$$

Indeed, we can rewrite $G_n(p, q)$ as

$$\sum_{\mathbf{k} \in \mathbb{N}^{p+1}} \sum_{\mathbf{j} \in \mathbb{N}^q} \frac{1}{s_1 s_2 \cdots s_p s_{p+1}^{n-1} \sigma_1 \sigma_2 \cdots \sigma_q (s_{p+1} + \sigma_q)},$$

where $s_\ell = k_1 + k_2 + \cdots + k_\ell$ and $\sigma_m = j_1 + j_2 + \cdots + j_m$. The analogous Drinfeld integral of $G_n(p, q)$ is given by

$$\int_D \prod_{j=1}^{p+1} \frac{dt_j}{1 - t_j} \prod_{k=p+2}^{n+p-1} \frac{dt_k}{t_k} \prod_{\ell=1}^{q} \frac{du_\ell}{1 - u_\ell} \frac{dt_{n+p}}{t_{n+p}},$$

where D is a domain in $\mathbb{R}^{n+p+q}$ defined as

$$0 < t_1 < t_2 < \cdots < t_{n+p}, \quad 0 < u_1 < u_2 < \cdots < u_q < t_{n+p}.$$

Therefore we have

$$G_n(p, q) = \frac{1}{p!q!(n-2)!} \int_{0 < t_1 < t_2 < 1} \left(\log \frac{1}{1 - t_1} \right)^p \left(\log \frac{1}{1 - t_2} \right)^q$$
$$\times \left(\log \frac{t_2}{t_1} \right)^{n-2} \frac{dt_1 dt_2}{(1 - t_1)t_2}.$$

Now we consider the double integral

$$\frac{1}{p!q!(k-2)!} \int_0^1 \int_0^1 \left(\log \frac{1}{1-t} \right)^p \left(\log \frac{u}{t} \right)^{k-2} \left(\log \frac{1}{1-u} \right)^q \frac{dt du}{tu}.$$

which is just the integral in (B.4.1) under the change of variables

$$(t, u) \mapsto (1 - t, 1 - u).$$

So its value is

$$\sum_{j=0}^{k-2}(-1)^j \zeta(\{1\}^{p-1}, k - j)\zeta(\{1\}^{q-1}, j + 2).$$

Next we evaluate the integration over the simplex $0 < t < u < 1$. Rewrite the integral as

$$\frac{1}{(p-1)!q!(k-1)!} \int_{0<t<u<1} \left(\log\frac{1}{1-t}\right)^{p-1} \left(\log\frac{u}{t}\right)^{k-1} \left(\log\frac{1}{1-u}\right)^q \frac{dt\,du}{(1-t)u},$$

so it is equal to $G_{k+1}(p - 1, q)$. In a similar way, the integration over the simplex $0 < u < t < 1$ is $(-1)^k G_{k+1}(q - 1, p)$.

Proposition B.4.1. *For positive integers k, p and q with $k \geq 2$, we have*

$$G_{k+1}(p - 1, q) + (-1)^k G_{k+1}(q - 1, p)$$
$$= \sum_{j=0}^{k-2}(-1)^j \zeta(\{1\}^{p-1}, k - j + 2)\zeta(\{1\}^{q-1}, j + 2).$$

Remark B.4.2. The above proposition is first obtained in [30] via a total different way.

Here we mention more useful properties of $G_n(p, q)$ from [30].

1. For any positive integer m, we have

$$\sum_{p+q=m} (-1)^p G_n(p, q) = \zeta(m + n).$$

It follows from that

$$\sum_{p+q=m} (-1)^p G_n(p, q)$$
$$= \sum_{p+q=m} \frac{(-1)^p}{p!q!(n-2)!} \int_{0<t_1<t_2<1} \left(\log\frac{1}{1-t_1}\right)^p \left(\log\frac{1}{1-t_2}\right)^q$$
$$\times \left(\log\frac{t_2}{t_1}\right)^{n-2} \frac{dt_1\,dt_2}{(1-t_1)t_2}$$

$$= \frac{1}{m!(n-2)!} \int_{0<t_1<t_2<1} \left(\log\frac{1-t_1}{1-t_2}\right)^m \left(\log\frac{t_2}{t_1}\right)^{n-2} \frac{dt_1 dt_2}{(1-t_1)t_2}$$

$$= \sum_{|\alpha|=m+n-1} \zeta(\alpha_1, \alpha_2, \ldots, \alpha_{m+1}+1)$$

$$= \zeta(m+n).$$

Such a property along with the evaluations of $G_n(p,q)+(-1)^{n+1}G_n(p-1,q+1)$ give the evaluation of $\zeta(\{1\}^{m-1}, n+1)$ when $m+n$ is even.

2. Apply simple transforms to the integral representation of $G_n(p,q)$ lead to the decomposition of $G_n(p,q)$ into sums of multiple zeta values as follows:

$$G_n(p,q) = \sum_{q_1+q_2=q} \binom{p+q_1}{p} \sum_{|\alpha|=n+q_2-1} \zeta(\{1\}^{p+q_1}, \alpha_1, \ldots, \alpha_{q_2+1}+1)$$

$$= \sum_{q_1+q_2=q} \binom{p+q_1}{p} \sum_{|\beta|=n+q_2-1} \zeta(\beta_1, \ldots, \beta_{n-1}+p+q_1+1)$$

$$= \sum_{q_1=0}^{q} \binom{p+q_1}{p} \sum_{|c|=p+q+1} \zeta(c_1, c_2, \ldots, c_{p+q_1+1}+n-1).$$

In the mean time, we also obtain the restricted sum formula

$$\sum_{|\alpha|=m} \zeta(\{1\}^p, \alpha_1, \alpha_2, \ldots, \alpha_r+1)$$

$$= \sum_{|c|=p+q} \zeta(c_1, c_2, \ldots, c_{p+1}+(m-r)+1).$$

The first decomposition follows from replacing the factor

$$\frac{1}{q!}\left(\log\frac{1}{1-t_2}\right)^q$$

in the integral representation of $G_n(p,q)$ by its binomial expansion

$$\sum_{q_1+q_2=q} \frac{1}{q_1!q_2!}\left(\log\frac{1}{1-t_1}\right)^{q_1}\left(\log\frac{1-t_1}{1-t_2}\right)^{q_2}.$$

Continue with a change of variable $(t_1, t_2) \mapsto (1-u_2, 1-u_1)$ with $0 < u_1 < u_2 < 1$ then leads to the second decomposition of $G_n(p,q)$.

To get the third decomposition, we employ Ohno's duality theorem and sum formula to get

$$\sum_{|\beta|=n+q_2-1} \zeta(\beta_1, \beta_2, \ldots, \beta_{n-1}+p+q_1+1)$$

$$= \sum_{|c|=p+q+1} \zeta(c_1, c_2, \ldots, c_{p+q_1+1}+n-1).$$

The third decomposition of $G_n(p,q)$ was used to determine the value of $\zeta(\{1\}^{m-1}, n+1)$ when $m+n$ is odd.

Appendix C

The Density of a Quadratic Form on Cayley Numbers

C.1. Preliminaries and Notations

Let $\mathbb{Z}$, $\mathbb{Q}$, $\mathbb{R}$, $\mathbb{C}$ denote the ring of integers, the fields of rational numbers, real numbers, and complex numbers, respectively. For a prime p, let $\mathbb{Z}_p$ and $\mathbb{Q}_p$ denote the ring of p-adic integers and the field of p-adic numbers, respectively. Let $\tau_p(a) = \max\{u \mid a/p^u \in \mathbb{Z}\}$ and $\overline{\tau}_p^{\nu}(a) = \min\{\nu, \tau_p(a)\}$. If R is a ring, we use $M_{m \times n}(R)$ to denote the ring of $m \times n$ matrices over R, and identify it with R^{mn} as usual. Further let $\wp_n$ be the set of $n \times n$ positive definite symmetric matrices over $\mathbb{R}$, $\mathfrak{H}_n$ the Siegel upper half-plane of degree n, and $\mathrm{Sp}(n, \mathbb{R})$ the real symplectic group of degree n. Specifically,

$$\mathfrak{H}_n = \left\{ X + iY \mid {}^t X = X \in M_n(\mathbb{R}), Y \in \wp_n \right\}.$$

If $S \in \wp_m \cap M_m(\mathbb{Z})$ satisfies

(a) S is unimodular, i.e., $\det(S) = 1$, and

(b) S is even, i.e., the diagonal entries are even integers,

251

then m must be a multiple of 8. Therefore the theta series $\vartheta_S(Z) = \sum_{G \in M_{m \times n}(\mathbb{Z})} e^{\pi i (S[G], Z)}$ is a modular form of weight $m/2$ on $\mathfrak{H}_n$, where $Z \in \mathfrak{H}_n$, (A, B) denotes the trace of AB, and $S[G] = {}^t G S G$.

Let S be such a matrix and $T \in \wp_n \cap M_n(\mathbb{Z})$, we use $A(S, T)$ to denote the number of solutions $G \in M_{m \times n}(\mathbb{Z})$ to the equation $S[G] = T$, and for a positive integer q, we use $A_q(S, T)$ to denote the number of incongruent integral solutions to the congruence $S[G] \equiv T \pmod{q}$. For each prime number p, the number

$$d_p(S, T) = \lim_{a \to \infty} p^{-a(mn - n(n+1)/2)} A_{p^a}(S, T)$$

is called the density at finite place p. Let $\mathfrak{N}$ be a relatively compact neighborhood of T in $\wp_n$, we define a map φ from $\mathbb{R}^{mn}$ to $\mathbb{R}^{n(n+1)/2}$ by $\varphi(G) = S[G]$. It is easy to see that $\varphi^{-1}(\mathfrak{N})$ has finite volume. Let

$$A_\infty(S, \mathfrak{N}) = \frac{\text{vol}(\varphi^{-1}(\mathfrak{N}))}{\text{vol}(\mathfrak{N})} \quad \text{and} \quad A_\infty(S, T) = \lim_{\mathfrak{N} \to T} A_\infty(S, \mathfrak{N}).$$

The limit exists when $\mathfrak{N}$ runs through a sequence of neighborhoods shrinking to T. We call it the density at infinite place.

Theorem C.1.1 (Siegelsche-Hauptsatz). [6, 49] *Notations as given above. We have*

$$A_\infty(S, T) \prod_p d_p(S, T) = \frac{\sum_{i=1}^h A(S_i, T)/A(S_i, S_i)}{\sum_{i=1}^h A(S_i, S_i)^{-1}},$$

where $S_1, S_2, \ldots, S_h$ *form a complete set of representatives of equivalent classes in the genus of* S.

Theorem C.1.2. [31, 51] *Suppose that* $E_k(Z)$ *is the Eisenstein series of weight* $k = 4n$ *on* $\mathfrak{H}_n$, *then*

$$E_k(Z) = \frac{\sum_{i=1}^h \vartheta_{S_i}(Z)/A(S_i, S_i)}{\sum_{i=1}^h A(S_i, S_i)^{-1}}.$$

Remark C.1.3. From Theorems C.1.1 and C.1.2, it is easy to see that the Fourier coefficient $a_k(T)$ equals $A_\infty(S, T) \prod_p d_p(S, T)$.

We now turn to Baily's results on the exceptional domain. For a field $\mathfrak{f}$, the Cayley numbers over $\mathfrak{f}$ is an eight-dimensional vector space $\mathscr{C}_\mathfrak{f}$ over $\mathfrak{f}$ with a basis $\mathfrak{B} = \{e_0, e_1, \ldots, e_7\}$ which is characterized by the following rules for multiplication [14]:

(i) $xe_0 = e_0x = x$ for all $x \in \mathscr{C}_f$;

(ii) $e_j^2 = -e_0$, $j = 1, 2, \ldots, 7$;

(iii) $e_1e_2e_4 = e_2e_3e_5 = e_3e_4e_6 = e_4e_5e_7 = e_5e_6e_1 = e_6e_7e_2 = e_7e_1e_3 = -e_0$.

For $x = \sum_{j=0}^{7} x_je_j \in \mathscr{C}_f$, let

(a) $\overline{x} = x_0e_0 - \sum_{j=1}^{7} x_je_j$;

(b) $N(x) = x\overline{x} = \sum_{j=0}^{7} x_j^2$, this function is called the norm of x;

(c) $T(x) = x + \overline{x} = 2x_0$, this function is called the trace of x;

(d) $\sigma(x, y) = 2\sum_{j=0}^{7} x_jy_j$ if $y = \sum_{j=0}^{7} y_je_j \in \mathscr{C}_f$.

Note that $N(x + y) = N(x) + N(y) + \sigma(x, y)$ and $\sigma(x, y) = T(x\overline{y}) = T(\overline{x}y)$. The ring $\mathcal{O}$ of integral Cayley numbers is a $\mathbb{Z}$-module in $\mathscr{C}_\mathbb{Q}$. It is generated by the $\mathbb{Z}$-basis $\mathcal{B}' = \{\alpha_0, \alpha_1, \ldots, \alpha_7\}$, where α_j are given as follows:

$$\alpha_0 = e_0, \quad \alpha_1 = e_1, \quad \alpha_2 = e_2, \quad \alpha_3 = -e_4,$$

$$\alpha_4 = \frac{1}{2}(e_1 + e_2 + e_3 - e_4), \quad \alpha_5 = \frac{1}{2}(-e_0 - e_1 - e_4 + e_5),$$

$$\alpha_6 = \frac{1}{2}(-e_0 + e_1 - e_2 + e_6), \quad \alpha_7 = \frac{1}{2}(-e_0 + e_2 + e_4 + e_7).$$

The $\mathbb{Z}$-module $\mathcal{O}$ is characterized as the maximal $\mathbb{Z}$-module in $\mathscr{C}_\mathbb{Q}$ with integral trace and norm [14]. We call an element w in $\mathcal{O}$ primitive modulo p if there does not exist $w' \in \mathcal{O}$ such that $w = pw'$. Let $\mathcal{J}$ be the set of 3×3 Hermitian matrices over $\mathscr{C}_\mathbb{R}$, specifically, $\mathcal{J}$ consists of matrices of the form

$$X = \begin{pmatrix} \xi_1 & x_{12} & x_{13} \\ \overline{x}_{12} & \xi_2 & x_{23} \\ \overline{x}_{13} & \overline{x}_{23} & \xi_3 \end{pmatrix}, \quad \xi_1, \xi_2, \xi_3 \in \mathbb{R}, \quad x_{12}, x_{13}, x_{23} \in \mathscr{C}_\mathbb{R}.$$

There is a product defined on $\mathcal{J}$: $X \circ Y = (XY + YX)/2$. For $X \in \mathcal{J}$,

$$\det(X) = \xi_1\xi_2\xi_3 - \xi_1N(x_{23}) - \xi_2N(x_{13}) - \xi_3N(x_{12}) + T((x_{12}x_{23})\overline{x}_{13}).$$

There is also an inner product on $\mathcal{J}$ defined by $(X, Y) = \mathrm{tr}(()X \circ Y)$. Finally, let $\mathcal{K}$ be the set of square elements $X \circ X$ of $\mathcal{J}$ and $\mathcal{K}^+$ the interior of $\mathcal{K}$. The exceptional domain of dimension 27 is contained in $\mathbb{C}^{27}$ and defined by

$$\mathcal{H}_3 = \{Z = X + iY \mid X \in \mathcal{J}, Y \in \mathcal{K}^+\}.$$

The lattice for our Fourier expansions are

$$\Lambda_3 = \mathcal{K} \cap M_3(\mathcal{O}), \quad \Lambda_i = \left\{ T \in M_i(\mathcal{O}) \;\middle|\; \begin{pmatrix} T & 0 \\ 0 & 0 \end{pmatrix} \in \Lambda_3 \right\} \quad \text{for } i = 1, 2.$$

Let $\Lambda_i / p^\nu \Lambda_i$ be the set of equivalent classes of the relation on Λ_i which asserts x equivalent to y if $x - y \in p^{nu} \Lambda_i$. Set

$$\mathcal{K}_3 = \{T \in \mathcal{K} \mid \det(T) \neq 0\},$$

$$\mathcal{K}_2 = \left\{ T = \begin{pmatrix} n & t & 0 \\ \bar{t} & m & 0 \\ 0 & 0 & 0 \end{pmatrix} \;\middle|\; n, m \in \mathbb{R}, n, m \geq 0, t \in \mathscr{C}_\mathbb{R}, nm - N(t) \neq 0 \right\},$$

$$\mathcal{K}_1 = \left\{ T = \begin{pmatrix} n & 0 & 0 \\ 0 & 0 & 0 \\ 0 & 0 & 0 \end{pmatrix} \;\middle|\; n \in \mathbb{R}, n \geq 0 \right\}.$$

There is an obvious injection from Λ_i to $\mathcal{K}_i$. When we mention a function of $T \in \Lambda_i$ (e.g., density), we always mean the function of the image of T in $\mathcal{K}_i$. The singular series on $\mathcal{H}_i$ are defined as

$$S^{(i)}(T, k) = \sum_{R \in \Lambda_i(\mathbb{Q})/\Lambda_i} \epsilon(R, T) \kappa(R)^{-k}$$

and

$$S_p^{(i)}(T, k) = \sum_{R \in \Lambda_i(\mathbb{Q}_p)/\Lambda_i(\mathbb{Z}_p)} \epsilon(R, T) \kappa_p(R)^{-k},$$

where $T \in \Lambda_i$, $\Lambda_i(\mathfrak{f}) = \Lambda_i \otimes_{\mathbb{Z}} \mathfrak{f}$, $\epsilon(R, T) = e^{2\pi i(R,T)}$, κ and κ_p are as defined in [5].

Theorem C.1.4. [5] *For the cases when* $\mathrm{rank}\, T = i$, *the Fourier coefficient* $a_k(T)$ *of the Eisenstein series* E_k *of weight* $k = 18n$ *is given by*

$$a_k(T) = \Gamma_k^{(i)} S^{(i)}(T, k) = \Gamma_k^{(i)} \prod_p S_p^{(i)}(T, k),$$

where

$$\Gamma_k^{(i)} = 2^{ki} \pi^{ki-2i(i-1)} \prod_{j=0}^{i-1} \Gamma(k - 4j)^{-1}$$

and $\Gamma(\cdot)$ is the usual gamma function. Furthermore, the singular series $S_p^{(i)}$ are indeed finite sums. We need only sum over those $R \in \Lambda_i(\mathbb{Q}_p)/\Lambda_i(\mathbb{Z}_p)$ with $\tau_p(R) > \tau_p(T')$.

We want to seek certain equation such that S_p equals d_p and the gamma factor Γ_k equals d_∞.

C.2. Interpretations as Densities

In view of the fact that the theta series $\sum_{h \in \mathcal{O}^2} e^{2\pi i(h\,{}^t\overline{h}, Z)}$ is a modular form of weight 4 on $\mathcal{H}_2$ [21], it is plausible that the expression $h_1\,{}^t\overline{h}_1 + h_2\,{}^t\overline{h}_2 + \cdots + h_n\,{}^t\overline{h}_n$, $h_i \in \mathcal{O}^3$ is the counterpart of the quadratic form on Siegel upper half-plane. More precisely, we will prove that the p-adic density of the equation

$$h_1\,{}^t\overline{h}_1 + h_2\,{}^t\overline{h}_2 + \cdots + h_n\,{}^t\overline{h}_n = T \tag{†}$$

is the p-adic factor $S_p^{(i)}(T, 4n)$ of the Euler product expansion of the Fourier coefficient of the Eisenstein series on $\mathcal{H}_3$ when $\operatorname{rank} T = i \le 2$. We still have some problems to handle the case when $\operatorname{rank} T = 3$.

By p-adic density of T, we roughly mean the ratio of the number of solutions to (†) in p-adic sense to the average number of solutions. There are $p^{27\nu}$ values of T modulo p^ν, each of them should be assumed $p^{24n\nu-27\nu}$ times in average. However when $T \in \Lambda_2$, the average number should be defined differently since (†) then reduces to $h_1\,{}^t\overline{h}_1 + h_2\,{}^t\overline{h}_2 + \cdots + h_n\,{}^t\overline{h}_n = T$, $h_i \in \mathcal{O}^2$. In this case we have only $p^{10\nu}$ choices of T, each is assumed $p^{16n\nu-10\nu}$ times in average. The idea to deal with $T \in \mathcal{K}_1$ is similar. We make these ideas precise in the following definition.

Definition C.2.1. The p-adic density $d_p(T, 4n)$ of $T \in \mathcal{K}_i$ is defined as

$$\lim_{\nu \to \infty} C_\nu(T, 4n),$$

where $C_\nu(T, 4n) = p^{-k} \cdot \#\{(h_1, \ldots, h_n) \mid h_1\,{}^t\overline{h}_1 + \cdots + h_n\,{}^t\overline{h}_n \equiv T \pmod{p^\nu}\}$ and $k = 8ni\nu - k_i\nu$ with $k_1 = 1$, $k_2 = 10$ and $k_3 = 27$.

The following lemmas show the relation between our equation and the contents of matrices which are used to define the singular series in Section 1. Here we abbreviate $e^{2\pi i x}$ as $\epsilon(x)$.

Lemma C.2.2. *For any integer u, one has*

$$\sum_{h \in \mathcal{O}/p^\nu \mathcal{O}} \epsilon\left(\frac{uN(h)}{p^\nu}\right) = p^{4\nu + 4\overline{\tau}_p^\nu(u)}.$$

Proof. If $h \in \mathcal{O}$, $h = \sum_{i=0}^{7} x_i \alpha_i$, then a direct computation gives $2N(h) = S[x]$, where $x = {}^t(x_0, x_1, \ldots, x_7)$, and

$$S = \begin{pmatrix} 2E_4 & B \\ -{}^t B & 2E_4 \end{pmatrix} \quad \text{with} \quad B = \begin{pmatrix} 0 & -1 & -1 & -1 \\ 1 & -1 & 1 & 0 \\ 1 & 0 & -1 & 1 \\ 1 & 1 & 0 & -1 \end{pmatrix}.$$

In p-adic sense, $S = \left(\begin{smallmatrix} 0 & E_4 \\ E_4 & 0 \end{smallmatrix}\right)$, so if $\tau_p(u) \geq \nu$, then $\overline{\tau}_p^\nu(u) = \nu$, and

$$\sum_{t \in \mathcal{O}/p^\nu \mathcal{O}} \epsilon\left(\frac{uN(t)}{p^\nu}\right) = \sum_{t \in \mathcal{O}/p^\nu \mathcal{O}} 1 = p^{8\nu} = p^{4\nu + 4\overline{\tau}_p^\nu(u)}.$$

Otherwise, we have

$$\sum_{t \in \mathcal{O}/p^\nu \mathcal{O}} \epsilon\left(\frac{uN(t)}{p^\nu}\right) = \sum_{x \in \mathbb{Z}^8/p^\nu \mathbb{Z}^8} \exp\left(\frac{\pi i u S[x]}{p^\nu}\right)$$

$$= \sum_{\substack{\lambda_j = 1 \\ j = 0,1,\ldots,7}}^{p^\nu} \epsilon\left(\frac{(\lambda_0\lambda_4 + \lambda_1\lambda_5 + \lambda_2\lambda_6 + \lambda_3\lambda_7)u}{p^\nu}\right)$$

$$= \prod_{j=0}^{3} \sum_{\lambda_j=1}^{p^\nu} \sum_{\lambda_{j+4}=1}^{p^\nu} \epsilon\left(\frac{(\lambda_j\lambda_{j+4})u}{p^\nu}\right)$$

$$= \prod_{j=0}^{3} \left(p^\nu \cdot \#\{\lambda_j \mid 1 \leq \lambda_j \leq p^\nu, u\lambda_j \equiv 0 \pmod{p^\nu}\}\right)$$

$$= \prod_{j=0}^{3} p^\nu p^{\overline{\tau}_p^\nu(u)} = p^{4\nu + 4\overline{\tau}_p^\nu(u)}.$$

$\square$

Lemma C.2.3. *For $r \in \Lambda_1(\mathbb{Q}) = \mathbb{Q}$ such that $p^\nu r \in \Lambda_1(\mathbb{Z}_p)$, we have*

$$\sum_{h \in \mathcal{O}/p^\nu \mathcal{O}} \epsilon(rN(h)) = p^{8\nu} \kappa_p(r)^{-4}.$$

Proof. Let $r = p^{-k}u$, u a p-adic unit and $k \leq \nu$, then we have $\kappa_p(r) = p^k$. By the previous lemma,

$$\sum_{h \in \mathcal{O}/p^\nu \mathcal{O}} \epsilon(rN(h)) = \sum_{h \in \mathcal{O}/p^\nu \mathcal{O}} \epsilon\left(\frac{uN(h)}{p^k}\right) = \sum_{h \in \mathcal{O}/p^\nu \mathcal{O}} \epsilon\left(\frac{up^{\nu-k}N(h)}{p^\nu}\right)$$
$$= p^{4\nu+4(\nu-k)} = p^{8\nu}\kappa_p(r)^{-4}.$$

$\square$

Lemma C.2.4. *For $R \in \Lambda_2(\mathbb{Q})$ such that $p^\nu R \in \Lambda_2(\mathbb{Z}_p)$, we have*

$$\sum_{h \in (\mathcal{O}/p^\nu \mathcal{O})^2} \epsilon((h^t\overline{h}, R)) = p^{16\nu}\kappa_p(R)^{-4}.$$

Proof. For $U = \left(\begin{smallmatrix}1 & t\\0 & 1\end{smallmatrix}\right)$, $\left(\begin{smallmatrix}1 & 0\\t & 1\end{smallmatrix}\right)$ or $\left(\begin{smallmatrix}0 & 1\\1 & 0\end{smallmatrix}\right)$, $t \in \mathcal{O}$, we have

$$\sum_{h \in (\mathcal{O}/p^\nu \mathcal{O})^2} \epsilon((h^t\overline{h}, R[U])) = \sum_{h \in (\mathcal{O}/p^\nu \mathcal{O})^2} \epsilon(((Uh)(^t\overline{h}{}^t\overline{U}), R)),$$

so it suffices to prove the lemma for diagonal matrices since we can diagonalize R using the three types of matrices. Let $R = \mathrm{diag}\,[\mu_1 p^{\nu_1}, \mu_2 p^{\nu_2}]$, $\nu_1 \leq \nu_2$ where μ_1 and μ_2 are p-adic units. Then

$$\sum_{h \in (\mathcal{O}/p^\nu \mathcal{O})^2} \epsilon((h^t\overline{h}, R)) = \sum_{\alpha \in \mathcal{O}/p^\nu \mathcal{O}} \epsilon(\mu_1 N(\alpha)p^{\nu_1}) \sum_{\beta \in \mathcal{O}/p^\nu \mathcal{O}} \epsilon(\mu_2 N(\beta)p^{\nu_2}),$$

so we have the following cases:

1. $\nu_1, \nu_2 \geq 0$. It follows $\kappa_p(R) = 1$ and the sum is equal to $p^{16\nu}$.

2. $\nu_1 < 0 \leq \nu_2$. We have $\kappa_p(R) = p^{-\nu_1}$ and the sum is equal to $p^{8\nu+4\nu_1}p^{8\nu}$, or $p^{16\nu}\kappa_p(R)^{-4}$.

3. $\nu_1 \leq \nu_2 < 0$. This implies $\kappa_p(R) = p^{-(\nu_1+\nu_2)}$ and the sum is equal to $p^{8\nu+4\nu_1}p^{8\nu+4\nu_2}$, or $p^{16\nu}\kappa_p(R)^{-4}$.

$\square$

Remark C.2.5. The identity $(h^t \overline{h}, R[U]) = ((Uh)(^t \overline{h}^t \overline{U}), R)$ reduces our problem to the case of diagonalized matrices here and in the succeeding sections. For example, if we diagonalize T in the congruence

$$h_1{}^t\overline{h}_1 + h_2{}^t\overline{h}_2 + \cdots + h_n{}^t\overline{h}_n \equiv T \pmod{p^\nu},$$

then

$$(h_1{}^t\overline{h}_1)[U] + (h_2{}^t\overline{h}_2)[U] + \cdots + (h_n{}^t\overline{h}_n)[U] \equiv T[U] \pmod{p^\nu}$$

and

$$(Uh_1)({}^t\overline{h}_1{}^t\overline{U}) + (Uh_2)({}^t\overline{h}_2{}^t\overline{U}) + \cdots + (Uh_n)({}^t\overline{h}_n{}^t\overline{U}) \equiv \mathrm{diag}\,[t_1, t_2] \pmod{p^\nu}.$$

So the number of representations of T is the same as one of its diagonalized matrices since the multiplication by U induces an isomorphism. We will not repeat this kind of argument in succeeding discussions.

To simplify the notations, we denote the frequently encountered expression

$$\sum_{h \in (\mathcal{O}/p^\nu \mathcal{O})^i} \epsilon\left(\frac{(h^t\overline{h}, R)}{p^\nu}\right)$$

as $F_\nu^{(i)}(R)$ for $i = 1, 2, 3$. It is not hard to show that

$$F_\nu^{(i)}(pR) = p^{8i} F_{\nu-1}^{(i)}(R).$$

Theorem C.2.6. *Notations as defined above, we have*

$$d_p(T, 4n) = S_p^{(i)}(T, 4n)$$

for $T \in \mathcal{K}_1$, $\mathcal{K}_2$ and $i = \mathrm{rank}\,T$.

Proof. By the Siegel-Babylonian process, the number of solutions to (†) modulo p^ν is given by

$$p^{-k_i\nu} \sum_{R \in \Lambda_i/p^\nu \Lambda_i} \sum_{h_j \in (\mathcal{O}/p^\nu \mathcal{O})^i} \epsilon\left(\frac{(h_1{}^t\overline{h}_1 + \cdots + h_n{}^t\overline{h}_n - T, R)}{p^\nu}\right)$$

$$= p^{-k_i\nu} \sum_{R \in \Lambda_i/p^\nu \Lambda_i} \epsilon((T, Rp^{-\nu}))\left(F_\nu^{(i)}(R)\right)^n$$

$$= p^{-k_i\nu} \sum_{\substack{R'\in\Lambda_i(\mathbb{Q}_p)/\Lambda_i(\mathbb{Z}_p) \\ p^\nu R'\in\Lambda_i}} \epsilon((T,R'))\left(p^{8i\nu}\kappa_p(R')^{-4}\right)^n$$

$$= p^{8in\nu-k_i\nu} \sum_{\substack{R'\in\Lambda_i(\mathbb{Q}_p)/\Lambda_i(\mathbb{Z}_p) \\ p^\nu R'\in\Lambda_i}} \epsilon((T,R'))\kappa_p(R')^{-4n}$$

$$= p^{8in\nu-k_i\nu} S_p^{(i)}(T,4n).$$

Dividing the number of solutions to (†) modulo p^ν by $p^{8in\nu-k_i\nu}$, and passing to limit, we get the assertion. □

Remark C.2.7. If we prove $F_\nu^{(3)}(R) = p^{24\nu}\kappa_p(R)^{-4}$, then it is readily to see that the previous theorem holds for $T \in \mathcal{K}_3$.

We will calculate the finite densities explicitly in the next section, and they coincide with the known results. Now we define the infinite density $d_\infty(T)$ and will see that d_∞ equals the gamma factor hence complete the whole analogy. In the following definition, we identify $\mathcal{K}_i$ with $\mathbb{R}^{k_i}$ in the obvious way and elements in $\mathcal{O}^i$ with their coordinates in $(\mathbb{R}^8)^i$ with respect to $\mathfrak{B}$.

Definition C.2.8. The expression $\sum_{j=1}^n h_j{}^t\overline{h}_j$, $h_j \in \mathcal{O}^i$, induces a map $\varphi_i\colon \mathbb{R}^{8i} \to \mathbb{R}^{k_i}$. If $T \in \mathcal{K}_i$, $d_\infty(T)$ is defined as the limit of the ratio of volumes

$$\frac{\mathrm{vol}(\varphi_i^{-1}(\mathcal{N}))}{\mathrm{vol}(\mathcal{N})},$$

where $\mathcal{N}$ runs through a sequence of relatively compact neighborhoods of T in $\mathbb{R}^{k_i}$.

C.3. Densities at Finite Places

We calculate the finite densities for T in $\mathcal{K}_1$ and $\mathcal{K}_2$. Note that when $T \in \mathcal{K}_1$, the problem also reduces to finding the Fourier coefficients of the classical Eisenstein series on the upper half-plane, which is well-known. However, we calculate it with our interpretation and compare with the known result to get an interesting arithmetic result concerning integral Cayley numbers. This will be described in more detail in the next section.

Theorem C.3.1. *If $t = p^{\alpha}t'$ with $(t', p) = 1$, then one has*

$$d_p(t, 4n) = (1 - p^{-4n}) \sum_{h=0}^{\alpha} p^{(1-4n)h}.$$

Proof. If t is relatively prime to p, then

$$d_p(t) = p^{-8n+1}p^{-1} \sum_{h_i \in \mathcal{O}/p\mathcal{O}} \sum_{v=1}^{p} \epsilon\left(\frac{v(N(h_1) + \cdots + N(h_n) - t)}{p}\right)$$

$$= p^{-8n} \sum_{v=1}^{p} \epsilon\left(\frac{vt}{p}\right) \left(\sum_{h \in \mathcal{O}/p\mathcal{O}} \epsilon\left(\frac{vN(h)}{p}\right)\right)^n$$

$$= p^{-8n} \sum_{v=1}^{p} p^{4n+4n\bar{\tau}_p^1(u)} \epsilon\left(\frac{vt}{p}\right)$$

$$= p^{-8n}(-p^{4n} + p^{8n}) = 1 - p^{-4n}.$$

If $t = p^{\alpha}t'$ with $(t', p) = 1$, one needs only to choose $\nu = \alpha + 1$, so

$$d_p(t) = p^{-8n+\nu}p^{-\nu} \sum_{h_i \in \mathcal{O}/p\mathcal{O}} \sum_{v=1}^{p^{\nu}} \epsilon\left(\frac{v(N(h_1) + \cdots + N(h_n) - t)}{p^{\nu}}\right)$$

$$= p^{-8n} \sum_{v=1}^{p^{\nu}} \epsilon\left(\frac{vt'}{p^{\nu-\alpha}}\right) \left(\sum_{h \in \mathcal{O}/p^{\nu}\mathcal{O}} \epsilon\left(\frac{vN(h)}{p^{\nu}}\right)\right)^n$$

$$= (1 - p^{-4n}) \sum_{h=0}^{\alpha} p^{(1-4n)h}.$$

$\square$

Now we turn to the case $T \in \mathcal{K}_2$ after the following lemmas.

Lemma C.3.2. *For nonnegative integers ν, a, b, let $\mathcal{R}_{\nu}(a, b)$ be the cardinality of the set*

$$\left\{ R = \begin{pmatrix} u & w \\ \overline{w} & v \end{pmatrix} \in \Lambda_2/p^{\nu}\Lambda_2 \;\middle|\; \overline{\tau}_p^{\nu}(u) \geq a, \overline{\tau}_p^{\nu}(v) \geq b, \det(R) \equiv 0 \pmod{p^{\nu}} \right\},$$

then we have

$$\mathcal{R}_{\nu}(a, b) = p^{9\nu-a-b} \sum_{k=0}^{a+b-1} \frac{\varphi(p^k)}{p^{4k}} + p^{9\nu} \sum_{k=a+b}^{\nu} \frac{\varphi(p^k)}{p^{5k}},$$

where $\varphi(\cdot)$ is the Euler phi-function.

Proof. A standard argument shows

$$\mathcal{R}_\nu(a,b) = p^{-\nu} \sum_{R \in \Lambda_2 / p^\nu \Lambda_2} \sum_{h=1}^{p^\nu} \epsilon\left(\frac{h \det(R)}{p^\nu}\right)$$

and the result follows by evaluation of the Gaussian sum. □

Lemma C.3.3 (Reduction Formula). *For* $T = \begin{pmatrix} \mu_1 p^\alpha & 0 \\ 0 & \mu_2 p^\beta \end{pmatrix}$, *let* $C_\nu(T)$
be as in Definition C.2.1,

$$H_\nu(T) = p^{-8n\nu} \sum_{R \in \Lambda_2 / p^\nu \Lambda_2} p^{4n \overline{\tau}_p^\nu(\det(R))} \epsilon\left(\frac{-(T,R)}{p^\nu}\right)$$

and

$$G_\nu(T) = p^{-4n\nu}(1 - p^{-4n}) \sum_{\substack{R \in \Lambda_2 / p^\nu \Lambda_2 \\ \det(R) \equiv 0 \pmod{p^\nu}}} \epsilon\left(\frac{-(T,R)}{p^\nu}\right).$$

Then we have

(a) $C_\nu(T) = H_\nu(T) + \sum_{k=1}^{\nu-1} G_k(T) + C_1(T) - H_1(T)$.

(b) $d_p(T) \lim_{\nu \to \infty} C_\nu(T) = C_{\alpha+\beta+2}(T) = \sum_{k=1}^{\alpha+\beta+1} G_k(T) + C_1(T) - H_1(T)$.

Proof. If $R = \begin{pmatrix} u & w \\ \overline{w} & v \end{pmatrix}$ is primitive, i.e., $u \not\equiv 0 \pmod{p}$ or $v \not\equiv 0 \pmod{p}$,
or w is a primitive element in $\mathcal{O}$, then $\kappa_p(R/p^\nu) = p^{2\nu - \overline{\tau}_p(\det(R))}$. By
Lemma C.2.3, we have

$$F_\nu^{(2)}(R) = p^{16\nu} \kappa_p \left(\frac{R}{p^\nu}\right)^{-4} = p^{8\nu + 4\overline{\tau}_p(\det(R))}.$$

We split the Gaussian sum into two parts according to whether R is primitive or not. The contribution from primitive R is

$$p^{-16n\nu} \sum_{\substack{R \in \Lambda_2 / p^\nu \Lambda_2 \\ R:\text{primitive}}} \left(F_\nu^{(2)}(R)\right)^n \epsilon\left(\frac{-(R,T)}{p^\nu}\right)$$

$$= p^{-8n\nu} \sum_{\substack{R \in \Lambda_2 / p^\nu \Lambda_2 \\ R:\text{primitive}}} p^{4n \overline{\tau}_p(\det(R))} \epsilon\left(\frac{-(R,T)}{p^\nu}\right)$$

$$= H_\nu(T) - p^{-8n\nu} \sum_{R' \in \Lambda_2 / p^{\nu-1} \Lambda_2} p^{4n \overline{\tau}_p(\det(p^2 R'))} \epsilon\left(\frac{-(R',T)}{p^{\nu-1}}\right)$$

$$= H_\nu(T) - p^{-8n(\nu-1)} \sum_{\substack{R'\in\Lambda_2/p^{\nu-1}\Lambda_2 \\ \overline{\tau}_p^{\nu-1}(\det(R'))\leq\nu-2}} p^{4n\overline{\tau}_p^{\nu-1}(\det(R'))}\epsilon\left(\frac{-(R',T)}{p^{\nu-1}}\right)$$

$$- p^{-4n\nu} \sum_{\substack{R'\in\Lambda_2/p^{\nu-1}\Lambda_2 \\ \det(R')\equiv 0 \pmod{p^{\nu-1}}}} \epsilon\left(\frac{-(R',T)}{p^{\nu-1}}\right)$$

$$= H_\nu(T) - H_{\nu-1}(T) + G_{\nu-1}(T).$$

And the contribution from nonprimitive R is given by

$$p^{-16n\nu} \sum_{\substack{R\in\Lambda_2/p^{\nu}\Lambda_2 \\ R=pR'}} \left(F_\nu^{(2)}(R)\right)^n \epsilon\left(\frac{(R,T)}{p^\nu}\right)$$

$$= p^{-16n(\nu-1)} \sum_{R'\in\Lambda_2/p^{\nu-1}\Lambda_2} \left(F_\nu^{(2)}(R)\right)^n \epsilon\left(\frac{(R',T)}{p^{\nu-1}}\right)$$

$$= C_{\nu-1}(T).$$

The assertion (a) follows easily by induction, and (b) follows from the property of Gaussian sum that H_ν for $\nu \geq \alpha + \beta + 2$. Hence C_ν is a constant for $\nu \geq \alpha + \beta + 2$.   □

We have converted the problem of finding C_ν to the above Gaussian sums which are easier to figure out. Calculation of the Gaussian sums gives

Lemma C.3.4. *Notations as introduced in Lemma C.3.3. We have*

(a) $C_1(T) - H_1(T) = 1 - p^{-4n}$;

(b) $G_h(T) = (1 - p^{-4n})\left((1-p^{-1})p^{(5-4n)h}\sum_{k=0}^{h}p^{4k} + p^{-1}p^{(9-4n)h}\right)$ *if* $0 \leq h \leq \alpha$;

(c) $G_h(T) = (1 - p^{-1})(1 - p^{-4n})p^{(5-4n)h}\sum_{k=0}^{\alpha}p^{4k}$ *if* $\alpha + 1 \leq h \leq \beta$;

(d) $G_h(T) = (1 - p^{-4n})\left((1-p^{-1})p^{(5-4n)h}\sum_{k=0}^{\alpha+\beta-h}p^{4k} - p^{-1}p^{4(\alpha+\beta+1)}\right)$
$\times p^{(1-4n)h}$ *if* $\beta \leq h \leq \alpha + \beta$; *and*

(e) $G_{\alpha+\beta+1}(T) = -p^{-1}p^{(5-4n)(\alpha+\beta+1)}(1 - p^{-4n})$.

Combining everything together, we get

Theorem C.3.5. *For* $T = \begin{pmatrix} \mu_1 p^\alpha & 0 \\ 0 & \mu_2 p^\beta \end{pmatrix}$ *with* $\beta \leq \alpha$, *we have*

$$d_p(T) = (1 - p^{-4n})(1 - p^{4-4n}) \sum_{h=0}^{\alpha} p^{(5-4n)h} \sum_{k=0}^{h} p^{4k}$$

$$+ (1 - p^{-4n})(1 - p^{4-4n}) \sum_{h=\alpha+1}^{\beta} p^{(5-4n)h} \sum_{k=0}^{\alpha} p^{4k}$$

$$+ (1 - p^{-4n})(1 - p^{4-4n}) \sum_{h=\beta+1}^{\alpha+\beta} p^{(5-4n)h} \sum_{k=0}^{\alpha+\beta-h} p^{4k}.$$

Remark C.3.6. After a re-grouping of our expression for d_p, it coincides with the result of Karel [37]. We can verify that the Fourier coefficients satisfy the Maaß condition which means the coefficient with respect to an arbitrary T can be expressed as linear combination of the coefficients with respect to those T with $\alpha = 0$. So the Eisenstein series is indeed a Maaß form.

C.4. Densities at Infinite Places

In this section, we compute the infinite densities directly from the definition, and combine the finite densities to obtain the formula of the Fourier coefficients. To begin, we have to choose a suitable sequence of neighborhood shrinking to T. When $T \in \mathcal{K}_1$, the problem reduces to the map $\varphi_1 \colon \mathbb{R}^{8n} \to \mathbb{R}$ given by

$$(x_{10}, \ldots, x_{17}, \ldots, x_{n0}, \ldots, x_{n7}) \mapsto \sum_{i=1}^{n} N\left(\sum_{j=0}^{7} x_{ij}\alpha_j\right) = \frac{1}{2}\sum_{i=1}^{n} S[x_i],$$

where $x_i = {}^t(x_{i0}, \ldots, x_{i7})$, and S is the 8×8 matrix given in Lemma C.2.2. It is very natural to choose the interval $\mathcal{N}'_\epsilon = [t - \epsilon, t + \epsilon]$ and let $\epsilon \to 0$. But for the sake of symmetry, one sees immediately that the result is the same if one chooses $\mathcal{N}_\epsilon = [t, t + \epsilon]$ although $\mathcal{N}_\epsilon$ is not an open set at all. We use this kind of "neighborhood" in this case and the next. The correctness is easy to justify.

Theorem C.4.1. *For* $t \in \Lambda_1$, *we have*

$$d_\infty(t) = \frac{(2\pi)^{4n}}{(4n-1)!} t^{4n-1}.$$

Proof. Note that

$$\text{vol}(\mathcal{N}_\epsilon) = \epsilon$$

and

$$\text{vol}(\varphi_1^{-1}(\mathcal{N}_\epsilon)) = \text{vol}\left\{(x_1, x_2, \ldots, x_n) \in \mathbb{R}^{8n} \,\middle|\, t \le \frac{1}{2}\sum_{i=1}^{n} S[x_i] \le t + \epsilon\right\}$$

$$= \frac{\pi^{4n}(2t + 2\epsilon)^{4n}}{(4n)!} - \frac{\pi^{4n}(2t)^{4n}}{(4n)!}.$$

The result follows easily if we observe that the limit of the quotient is just the derivative of a function. $\square$

Combining the result of the previous section, we obtain the Fourier coefficient of the Eisenstein series for $T \in \mathcal{K}_1$ is given by

$$d_\infty(t) \prod_p d_p(t) = -\frac{8n}{B_{4n}} \sum_{d|t} d^{4n},$$

where B_{4n} is the Bernoulli number. This is not surprising at all since $\mathcal{H}_1$ is just the upper half-plane and the Siegelsche-Hauptsatz applies in this case. What is interesting is the arithmetic result behind it.

Corollary C.4.2. *We have*

$$\#\{t \in \mathcal{O} \mid N(t) = m\} = 240 \sum_{d|m} d^3.$$

Proof. By Siegelsche-Hauptsatz, the product of all densities is a weighted average of the number of solutions to h quadratic equations $S_i[x] = 2T$, where h is the number of equivalent classes in the genus of S. But in the case $n = 1$, there is only one equivalent class [47]. Hence we get the actual number of solutions in this case. $\square$

When $T \in \Lambda_2$, the problem reduces to the map $\varphi_2 \colon \mathbb{R}^{16n} \to \mathbb{R}^{10}$. Here we identify $T = \begin{pmatrix} t_1 & \sum_{i=0}^{7} a_i\alpha_i \\ \sum_{i=0}^{7} a_i\bar{\alpha}_i & t_2 \end{pmatrix} \in \Lambda_2$ with ${}^t(t_1, t_2, a_0, \ldots, a_7) \in \mathbb{R}^{10}$, then φ_2 is given by

$$(x_{10}, \ldots, x_{17}, y_{10}, \ldots, y_{17}, \ldots, x_{n0}, \ldots, x_{n7}, y_{n0}, \ldots, y_{n7})$$

$$\mapsto \begin{pmatrix} \sum_{i=1}^{n} N(x_i) & \sum_{i=1}^{n} x_i\bar{y}_i \\ \sum_{i=1}^{n} y_i\bar{x}_i & \sum_{i=1}^{n} N(y_i) \end{pmatrix},$$

where $x_i = \sum_{j=0}^{7} x_{ij}\alpha_j$ and $y_i = \sum_{j=0}^{7} y_{ij}\alpha_j$. It suffices to consider only diagonalized matrix $T = \left(\begin{smallmatrix} t_1 & 0 \\ 0 & t_2 \end{smallmatrix}\right)$ as in the finite case. Let

$$\mathcal{N}_\epsilon = \left\{(a_1, a_2, s_0, \ldots, s_7) \in \mathbb{R}^{10} \mid t_i \le a_i \le t_i + \epsilon, 0 \le s_i \le \epsilon^2 \right\}.$$

We can start our calculation after the following definition and interesting lemma which enable us to compute Cayley numbers with the help of matrices.

Definition C.4.3. For a Cayley number $x \in \mathscr{C}_\mathbb{R}$, let $\hat{x}$ and $\tilde{x}$ in $\mathbb{R}^8$ be its coordinates with respect to the ordered bases $\mathfrak{B} = \{e_0, e_1, \ldots, e_7\}$ and $\mathfrak{B}' = \{\alpha_0, \alpha_1, \ldots, \alpha_7\}$, respectively. Let A be the matrix of change of basis, i.e., $A\tilde{x} = \hat{x}$ for all $x \in \mathscr{C}_\mathbb{R}$. Let $S_0, S_1, \ldots, S_7$ be matrices such that $x\bar{y} = \sum_{i=0}^{7} {}^t(\hat{x}S_i\hat{y}_i)e_i$ and $S_0', S_1', \ldots, S_7'$ be matrices such that $x\bar{y} = \sum_{i=0}^{7} {}^t(\tilde{x}S_i\tilde{y}_i)\alpha_i$.

The following are several interesting properties about the matrices.

Lemma C.4.4.   (a) *If $S_0 = E_8$, the 8×8 identity matrix, then $S_i^2 = -S_0$ for $i = 1, 2, \ldots, 7$.*

(b) *The matrix S_iS_j is anti-symmetric for all $i \ne j$. In particular, S_i is anti-symmetric for $i = 1, 2, \ldots, 7$.*

(c) *For a column vector $v \in \mathbb{R}^8$, we have*

$$A \begin{pmatrix} {}^tvS_0' \\ \vdots \\ {}^tvS_7' \end{pmatrix} = \begin{pmatrix} {}^tv{}^tAS_0 \\ \vdots \\ {}^tv{}^tAS_7 \end{pmatrix} A.$$

(d) ${}^tAA = \frac{1}{2}(S_0' + {}^tS_0') = \frac{1}{2}S$ *and* $\det(A) = 1/2^4$.

(e) *For $x_i \in \mathscr{C}_\mathbb{R}$, let*

$$D = \begin{pmatrix} \hat{x}_1 S_0 & \cdots & \hat{x}_n S_0 \\ \vdots & \ddots & \vdots \\ \hat{x}_1 S_7 & \cdots & \hat{x}_n S_7 \end{pmatrix},$$

then $|\det(D^tD)| = \left(\sum_{i=1}^{n} N(x_i)\right)^8$.

Proof. It is easy to check (a), (b) and (d) by direct computations. The matrices involved can be written down explicitly. For (c), let x, $y \in \mathscr{C}_{\mathbb{R}}$ such that $\widetilde{x} = v$ and $\widetilde{y} = w$. It follows from Definition C.4.3 that $A\widetilde{x}\widetilde{y} = \widehat{x}\widetilde{y}$, so

$$
A \begin{pmatrix} {}^t v S_0' w \\ \vdots \\ {}^t v S_7' w \end{pmatrix} = A \begin{pmatrix} {}^t \widetilde{x} S_0' \widetilde{y} \\ \vdots \\ \widetilde{x} S_7' \widetilde{y} \end{pmatrix} = \begin{pmatrix} \widehat{x} S_0 \widehat{y} \\ \vdots \\ \widehat{x} S_7 \widehat{y} \end{pmatrix} = \begin{pmatrix} {}^t \widetilde{x}^t A S_0 A \widetilde{y} \\ \vdots \\ \widetilde{x}^t A S_7 \widetilde{y} \end{pmatrix}
$$

$$
= \begin{pmatrix} {}^t \widetilde{x}^t A S_0 \\ \vdots \\ \widetilde{x}^t A S_7 \end{pmatrix} A \widetilde{y} = \begin{pmatrix} {}^t \widetilde{x}^t A S_0 \\ \vdots \\ \widetilde{x}^t A S_7 \end{pmatrix} Aw,
$$

thus we conclude (c). Clearly, the matrix $D^t D$ is indeed a diagonal matrix. If $i \neq j$, then the (i,j)th entry is

$$
\sum_{k=1}^{n} {}^t\widehat{x}_k S_i{}^t S_j \widehat{x}_k = -\sum_{k=1}^{n} {}^t\widehat{x}_k S_i S_j \widetilde{x}_k = 0
$$

by (b). So the (i,i)th entry is

$$
\sum_{k=1}^{n} {}^t\widehat{x}_k S_i{}^t S_i \widehat{x}_k = \sum_{k=1}^{n} {}^t\widehat{x}_k \widehat{x}_k = \sum_{k=1}^{n} N(x_k). \qquad \square
$$

Theorem C.4.5. *For $T \in \Lambda_2$, we have*

$$
d_\infty(T) = \frac{(2\pi)^{8n-4}}{(4n-1)!(4n-5)!}(\det T)^{4n-5}.
$$

Proof. For the same reason as in the finite case, we can consider only diagonalized matrix $T = \begin{pmatrix} t_1 & 0 \\ 0 & t_2 \end{pmatrix}$. Let $\mathcal{N}_\epsilon$ be the neighborhood chosen in Definition C.4.3, then

$$
\text{vol}(\mathcal{N}_\epsilon) = \epsilon^{18}
$$

and

$$
\text{vol}(\varphi_2^{-1}(\mathcal{N}_\epsilon)) = \int_{R_1} d\widetilde{x}_1 \cdots d\widetilde{x}_n \int_{R_2} d\widetilde{y}_1 \cdots d\widetilde{y}_n, \qquad (\ddagger)
$$

where

$$
R_1 : 2t_1 \leq \sum_{i=1}^{n} {}^t\widetilde{x}_i S \widetilde{x}_i \leq 2t_1 + 2\epsilon
$$

and

$$R_2 : \begin{cases} 2t_2 \le \sum_{i=1}^{n} {}^t\widetilde{y}_i S \widetilde{y}_i \le 2t_2 + 2\epsilon; \\ 0 \le \sum_{j=1}^{n} {}^t\widetilde{x}_j S'_i \widetilde{y}_j \le \epsilon^2, \quad i = 0, 1, \dots, 7. \end{cases}$$

Make a change of variable $\widetilde{y}_i = A^{-1} y'_i$ for $i = 1, 2, \dots, n$. Then the Jacobian J_1 for this change of variable is 2^{4n}, and the conditions concerning the $\widetilde{y}'_i$'s are transformed into

$$t_2 \le \sum_{i=1}^{n} {}^t y'_i y'_i \le t_2 + \epsilon \quad \text{and} \quad 0 \le \sum_{j=1}^{n} {}^t\widetilde{x}_j S'_i A^{-1} y'_j \le \epsilon^2.$$

Geometrically, the region is a $(8n)$-dimensional ball intersected by 8 hyperplanes, so the volume is roughly the product of the volume of a $(8n-8)$-dimensional ball and the distances between the origin and the hyperplanes. We realize the idea through the following change of variable

$$\begin{pmatrix} y_1 \\ \vdots \\ y_n \end{pmatrix} = B \begin{pmatrix} y'_1 \\ \vdots \\ y'_n \end{pmatrix} = \begin{pmatrix} {}^t\widetilde{x}_1 S'_0 A^{-1} & \cdots & {}^t\widetilde{x}_n S'_0 A^{-1} \\ \vdots & \ddots & \vdots \\ {}^t\widetilde{x}_n S'_7 A^{-1} & \cdots & {}^t\widetilde{x}_n S'_7 A^{-1} \\ & B' & \end{pmatrix} \begin{pmatrix} y'_1 \\ \vdots \\ y'_n \end{pmatrix}$$

$$= \begin{pmatrix} D'' \\ B' \end{pmatrix} \begin{pmatrix} y'_1 \\ \vdots \\ y'_n \end{pmatrix}.$$

We ask the rows of B' constitute an orthonormal basis of the orthogonal complement in $\mathbb{R}^{8n}$ of the first 8 rows of B. The Jacobian J_2 of the second change of variables is

$$|\det(B^{-1})| = (\det(B^t B))^{-1/2} = (\det(D''^t D''))^{-1/2}.$$

As in the proof of Lemma C.4.4(e), we have

$$AD'' = \begin{pmatrix} {}^t\widetilde{x}_1 {}^t A S_0 & \cdots & {}^t\widetilde{x}_n {}^t A S_0 \\ \vdots & \ddots & \vdots \\ {}^t\widetilde{x}_n {}^t A S_7 & \cdots & {}^t\widetilde{x}_n {}^t A S_7 \end{pmatrix}$$

and

$$\det(AD''^t D''^t A) = \left(\sum_{i=1}^{n} {}^t\widetilde{x}_i {}^t AA \widetilde{x}_i \right)^8 = \left(\sum_{i=1}^{n} {}^t\widetilde{x}_i \frac{1}{2} S \widetilde{x}_i \right)^8 = \left(\sum_{i=1}^{n} N(x_i) \right)^8.$$

So $J_2 = 2^4 \left(\sum_{i=1}^n N(x_i) \right)^{-4}$ and it is clear that if we replace J_2 by $2^4 t_1^{-4}$, the limit of the quotient of volumes does not change. The conditions concerning the y_i''s are transformed into

$$0 \leq y_{1j} \leq \epsilon^2, \quad j = 0, 1, \ldots, 7,$$

where y_{1j} is the jth component of y_1, and

$$t_2 \leq ({}^tB^{-1}B^{-1}) \left[{}^t(y_1, y_2, \ldots, y_n) \right] \leq t_2 + \epsilon.$$

The terms in the last equation without y_1 can be represented by $B'^t B'$, which is a positive definite quadratic form with determinant 1. Putting all together, we have

$$\int_{R_2} d\widetilde{y}_1 \cdots d\widetilde{y}_n$$

$$= \int_{\substack{0 \leq y_{1j} \leq \epsilon^2 \\ 0 \leq j \leq 7}} dy_1 \int_{t_2 \leq ({}^tB^{-1}B^{-1})[{}^t(y_1,\ldots,y_n)] \leq t_2 + \epsilon} J_1 J_2 \, dy_2 \cdots dy_n$$

$$\approx \int_{\substack{0 \leq y_{1j} \leq \epsilon^2 \\ 0 \leq j \leq 7}} dy_1 \int_{t_2 \leq (B^t B)[{}^t(y_2,\ldots,y_n)] \leq t_2 + \epsilon} 2^{4(n-1)} t_1^{-4} \, dy_2 \cdots dy_n$$

$$= \epsilon^{16} t_1^{-4} 2^{4(n-1)} \left[\frac{\pi^{4n-4}(t_2 + \epsilon)^{4n-4}}{(4n-4)!} - \frac{\pi^{4n-4} t_2^{4n-4}}{(4n-4)!} \right].$$

We can take the limit for this part first since it does not involve the variables in the first iterated integral in (‡). The limit of the second integral to ϵ^{17} is readily seen to be

$$(2\pi)^{4n-4} \frac{t_1^{-4} t_2^{4n-5}}{(4n-5)!},$$

and the limit of the ratio of the first iterated integral to ϵ is

$$(2\pi)^{4n} \frac{t_1^{4n-1}}{(4n-1)!},$$

so they give

$$(2\pi)^{8n-4} \frac{t_1^{4n-5} t_2^{4n-5}}{(4n-1)!(4n-5)!} = (2\pi)^{8n-4} \frac{(\det(T))^{4n-5}}{(4n-1)!(4n-5)!}. \qquad \square$$

Remark C.4.6. This result also coincides with Karel [37]. They have an inessential error of a numerical factor which has been pointed out in [21].

Bibliography

[1] L. V. Ahlfors, *Complex Analysis: An Introduction to the Theory of Analytic Functions of One Complex Variable*, Third edition, McGraw-Hill, Inc., Auckland, 1979.

[2] A. N. Andrianov, *Quadratic Forms and Hecke Operators*, Springer-Verlag, Berlin, 1987.

[3] T. M. Apostol, *Introduction to Analytic Number Theory*, Springer-Verlag, New York, 1976.

[4] T. Arakawa and M. Kaneko, *Multiple zeta values, poly-Bernoulli numbers, and related zeta functions*, Nagoya Math. J. **153** (1999), 189–209.

[5] W. L. Baily, Jr., *An exceptional arithmetic group and its Eisenstein series*, Ann. of Math. (2) **91** (1970), 512–549.

[6] ———, *Introductory Lectures on Automorphic Forms*, Publications of the Math. Soc. of Japan, no. 12, Iwanami Shoten, Publishers and Princeton University Press, 1973.

[7] B. C. Berndt, *Ramanujan's Notebooks, Part I and II*, Springer-Verlag, New York, 1985, 1989.

[8] S. I. Borevič and I. R. Šafarevič, *Zahlentheorie*, Birkhäuser Verlag, 1966.

[9] J. M. Borwein and D. M. Bradley, *Thirty-two Goldbach variations*, Int. J. Number Theory **2** (2006), no. 1, 65–103.

269

[10] J. M. Borwein and R. Girgensohn, *Evaluation of triple Euler sums*, Electron. J. Combin. **3** (1996), no. 1, 1–27.

[11] D. Bowman and D. M. Bradley, *Multiple polylogarithms: a brief survey*, Contemp. Math. **291** (2001), 71–92.

[12] S.-T. Chang and M. Eie, *An arithmetic property of Fourier coefficients of singular modular forms on the exceptional domain*, Trans. Amer. Math. Soc. **353** (2000), 539–556.

[13] K.-W. Chen and M. Eie, *Explicit evaluations of extended Euler sums*, J. Number Theory **117** (2006), 31–52.

[14] H. S. M. Coxeter, *Integral Cayley numbers*, Duke Math. J. **13** (1946), 561–578.

[15] M. Eichler and D. Zagier, *The Theory of Jacobi Forms*, Birkhäuser, Boston, 1985.

[16] M. Eie, *A note on Bernoulli numbers and Shintani generalized Bernoulli polynomials*, Trans. Amer. Math. Soc. **348** (1996), no. 3, 1117–1136.

[17] ———, *Dimensions of spaces of Siegel cusp forms of degree two and three*, Mem. Amer. Math. Soc. **50** (1984), no. 304, 1–184.

[18] ———, *Fourier coefficients of Jacobi forms over Cayley numbers*, Rev. Mat. Iberoamericana **11** (1995), no. 1, 125–142.

[19] ———, *Jacobi-Eisenstein series of degree two over Cayley numbers*, Rev. Mat. Iberoamericana **16** (2000), no. 3, 571–596.

[20] ———, *The cohomology group associated with Jacobi cusp forms over Cayley numbers*, Amer. J. Math. **120** (1998), no. 4, 811–826.

[21] M. Eie and A. Krieg, *The Maaß space on the half-plane of Cayley numbers of degree two*, Math. Zeitschrift **210** (1992), 113–128.

[22] ———, *The theory of Jacobi forms over the Cayley numbers*, Trans. Amer. Math. Soc. **342** (1994), 793–805.

[23] M. Eie and W.-C. Liaw, *A restricted sum formula among multiple zeta values*, to appear in J. Number Theory (2008).

[24] ———, *Euler sums with Dirichlet characters*, Acta Arith. **130** (2007), no. 2, 99–125.

[25] M. Eie, W.-L. Liaw and Y. L. Ong, *Explicit evaluations of triple Euler sums*, to appear in Int. J. Number Theory (2008).

[26] M. Eie, W.-C. Liaw and F.-Y. Yang, *On evaluation of generalized Euler sums of even weight*, Int. J. Number Theory **1** (2005), no. 2, 225–242.

[27] M. Eie and Y. L. Ong, *A generalization of Kummer's congruences*, Abh. Math. Sem. Univ. Hamburg **67** (1997), 149–157.

[28] M. Eie, Y. L. Ong and F.-Y. Yang, *Evaluating double Euler sums over rationally deformed simplices*, Int. J. Number Theory **1** (2005), no. 3, 439–458.

[29] M. Eie and C.-S. Wei, *A short proof for the sum formula and its generalization*, submitted (2007).

[30] ———, *New Euler sums with two branches*, submitted (2008).

[31] E. Freitag, *Siegelsche Modulfunktionen*, Springer-Verlag, Berlin, 1983.

[32] A. Granville, *A decomposition of Riemann's zeta function*, Analytic number theory (Kyoto, 1996), 95–101.

[33] J. Hadamard, *Sur la distribution des zéros de la fonction $\zeta(s)$ et ses conséquences arithmétiques*, Bull. Soc. Math. France **24** (1896), 199–220.

[34] M. E. Hoffman, *Multiple harmonic series*, Pacific J. Math. **152** (1992), no. 2, 275–290.

[35] M. E. Hoffman and C. Moen, *Sums of triple harmonic series*, J. Number Theory **60** (1996), no. 2, 329–331.

[36] N. Jacobson, *Basic Algebra I*, Second edition, W. H. Freeman and Company, New York, 1985.

[37] M. Karel, *Fourier coefficients of certain Eisenstein series*, Ann. of Math. **99** (1974), 176–202.

[38] N. Koblitz, *p-Adic Analysis: a short course on recent work*, Cambridge University Press, 1980.

[39] _____, *p-Adic Numbers, p-Adic Analysis, and Zeta-Functions*, Second edition, Springer-Verlag, New York, 1984.

[40] J. E. Littlewood, *Sur la distribution des nombres premiers*, Comptes Rendus Acad. Sci. **158** (1914), 1869–1872.

[41] C. Markett, *Triple sums and the Riemann zeta function*, J. Number Theory **48** (1994), no. 2, 113–132.

[42] T. Miyake, *Modular Forms*, Springer-Verlag, New York, 1989.

[43] N. Nielsen, *Die Gammafunktion, Band I; Handbuch der Theorie der Gammafunktion, Band II; Theorie des Integrallogarithmus und verwandter Transzendenten*, 1906. Reprinted together as Die Gammafunktion, Chelsea Publishing Co., New York 1965.

[44] Y. Ohno, *A generalization of the duality and sum formulas on the multiple zeta values*, J. Number Theory **74** (1999), no. 1, 39–43.

[45] H. Rademacher, *Topics in Analytic Number Theory*, Springer-Verlag, New York, 1980.

[46] B. Schoeneberg, *Elliptic Modular Functions: an introduction*, Springer-Verlag, New York, 1974.

[47] J.-P. Serre, *A Course in Arithmetic*, Springer-Verlag, New York, 1973.

[48] G. Shimura, *Introduction to the Arithmetic Theory of Automorphic Functions*, Publications of the Math. Soc. of Japan **11**, Iwanami Shoten, Publishers and Princeton University Press, 1971.

[49] C. L. Siegel, *Über die analytische Theorie der quadratischen Formen*, Ann. of Math. **36** (1935), 527–606.

[50] Ch. de la Vallée Poussin, *Recherches analytiques sur la théorie des nombres premiers*, Ann. Soc. Sci. Bruxelles **20**$_2$ (1896), 183–256, 281–297.

[51] E. Witt, *Eine Identität zwischen Modulformen zweiten Grades*, Abh. Math. Sem. Hansisch. Univ. **14** (1941), 323–337.

[52] D. Zagier, *Values of zeta functions and their applications*, Progr. Math. **120** (1994), 497–512.

Index